Günter Scholz and Manuela Gehringer
Thermoplastic Elastomers

Also of Interest

Physical Chemistry of Polymers.
A Conceptual Introduction
Sebastian Seiffert, 2020
ISBN 978-3-11-067280-0, e-ISBN (PDF) 978-3-11-067281-7,
e-ISBN (EPUB) 978-3-11-067284-8

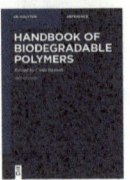

Handbook of Biodegradable Polymers
Catia Bastioli (Hrgs.), 2020
ISBN 978-1-5015-1921-5, e-ISBN (PDF) 978-1-5015-1196-7,
e-ISBN (EPUB) 978-1-5015-1198-1

Polyols for Polyurethanes.
Volume 1
Mihail Ionescu, 2019
ISBN 978-3-11-064033-5, e-ISBN (PDF) 978-3-11-064410-4,
e-ISBN (EPUB) 978-3-11-064051-9

Latex Dipping.
Science and Technology
David M. Hill, 2019
ISBN 978-3-11-063782-3, e-ISBN (PDF) 978-3-11-063809-7,
e-ISBN (EPUB) 978-3-11-063823-3

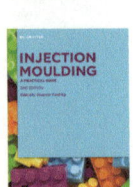

Injection Moulding.
A Practical Guide
Vannessa Goodship (Hrsg.), 2020
ISBN 978-3-11-065302-1, e-ISBN (PDF) 978-3-11-065481-3,
e-ISBN (EPUB) 978-3-11-065303-8

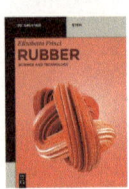

Rubber.
Science and Technology
Elisabetta Princi, 2019
ISBN 978-3-11-064031-1, e-ISBN (PDF) 978-3-11-064032-8,
e-ISBN (EPUB) 978-3-11-064052-6

Günter Scholz and Manuela Gehringer

Thermoplastic Elastomers

At a Glance

DE GRUYTER

Authors
Günter Scholz
formerly BASF Polyurethanes GmbH
R&D Elastomers
49448 Lemfoerde
guenter.alfreds@gmail.com

Manuela Gehringer
ALLOD Werkstoff GmbH & Co. KG
Steinacher Str. 3
91593 Burgbernheim
manuela.gehringer@allod.com

ISBN 978-3-11-073983-1
e-ISBN (PDF) 978-3-11-073984-8
e-ISBN (EPUB) 978-3-11-073998-5

Library of Congress Control Number: 2021936858

Bibliographic information published by the Deutsche Nationalbibliothek
The Deutsche Nationalbibliothek lists this publication in the Deutsche Nationalbibliografie;
detailed bibliographic data are available on the Internet at http://dnb.dnb.de.

—

Adults perceive what they believe to see.
Childs see what they observe.

Günter Alfred

Acknowledgments

We send many thanks for their helpful contributions to:
Stefan Zepnik (TPV)
Gert Joly (TPV)
Jörg Sänger (TPS)
Erik Licht (TPO)
Jürgen Hättig (TPU)
Freddy Gruber (TPC, TPA)
Markus Susoff (physical basics)
Stephanie Waschbüsch for continuous motivation and publication support
Felicia Bokel for creating the English version.

Many results had been taken from the Bachelor Theses of:
Nils Nagel
Jana Duhme

Many thanks to the BASF Polyurethanes GmbH to make it possible and providing the diagrams.

https://doi.org/10.1515/9783110739848-202

Prologue

Thermoplastic elastomers, briefly TPE, represent a unique product family in the world of plastic materials. They are young compared to most of the plastics and are a midsize player with a market volume of around 6 million metric tons in 2018. The tendency shows a continuous increase. They are highly versatile in both chemical construction and technical application, and the name describes the link between thermoplastic materials and rubber elastomers. Meanwhile, many abbreviations, names, and wordings circulate in the literature and public communication, and this handy publication with a basic level of scientific aspiration shall help the beginner in getting a fast and qualified introduction to these materials. The new aspect here is the deeper look into elastomeric properties, because TPE are originally created to become a thermoplastic alternative to rubber elastomers. Accordingly, the reader will gain a perspective look from every family about specific tensile–elongation measurements as well as dynamic mechanical analysis, which had been detected from a little collection of common standard products. Of course, this does not cover the full width of the material range of every TPE family entirely. When searching for the most suitable material for a technical solution, the counsel of the supplier is highly recommended.

Often people desire to compare materials, and they tend to take data out from the literature or to make comparable measurements. The biggest hurdle is the selection of certain grades from every TPE family. Despite such a comparison, it must be clear that a property profile of one material can never cover the complete profile of another one. Such investigations can only be made for certain applications finding a better technical solution. Some property comparison among TPE cannot always be avoided. We use experience and instinct to make comparisons where valid, but focus on one distinct chapter for every type of TPE.

This book is not a work of reference for product data and specific information needed for the work with a certain grade. For this, consultation with a material supplier is recommended. Our brief glance intends to spark interest, continuing further studies, and to help the user to better target and find the right material. Seeing the versatility of using TPE should be enjoyable. Having a basic understanding of plastic technology is very helpful for employing this book. It does not represent a comprehensive manual but shall be seen as a supplement and it has the intention to provide a quick and qualified insight.

It was not possible to write this book alone. We would like to express our gratitude to many experts from the TPE world who often gave an unrecognized contribution inciting thoughts and stimulating discussions.

Günter Scholz, Lemförde
May 2021

https://doi.org/10.1515/9783110739848-203

Contents

1 Introduction to TPE

Among the thermoplastic elastomers (TPE), inherently two different molecular structures exist; these are the compounds of polymeric materials TPO (thermoplastic polyolefin) and TPV (thermoplastic vulcanizate) and the amorphous or semicrystalline copolymers TPO (olefinic block copolymer), TPS (styrene block copolymers), TPU (urethane block copolymer), TPC (ester block copolymer), and TPA (amide block copolymer). It sounds strange that TPO is visible in both classes, but there is a project close to be finished in the ISO committee to clarify this ambiguity situation. A new draft standard is published, which is the normalized standard (norm) DIN EN ISO 18064, where even the general definition of TPE is disclosed. It is "a polymer, or blend of polymers, that has properties at its service temperature similar to those of vulcanized rubber but can be processed and reprocessed at an elevated temperature like a thermoplastic."

Even other polymer mixtures are possible to put under this definition. Such compounds are labeled as TPZ (elastomeric polymer blends). This part will not be introduced in this book because this would exist as a stand-alone topic. It is complex enough in TPS, TPO, and TPV chapters to discuss compounds because these materials are almost solely used as polymer blends in technical applications. Unless otherwise stated, properties and applications of TPS and TPV only include blends.

Considering the typical structure of a TPE copolymer with a crystalline or amorphous hard phase connected to a flexible soft phase, one could imagine an elastomer with a chemically bonded filler at nanoscale dimension (see Fig. 1.1). And this is, in fact, the truth. The "filler" melts at elevated temperature and the amorphous section softens like a glass. It can be interpreted as a physical network that is completely different from a chemically cross-linked vulcanized rubber.

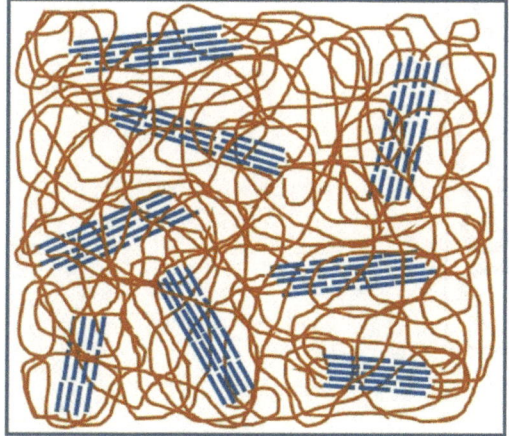

Fig. 1.1: Structure of a semicrystalline TPE.

https://doi.org/10.1515/9783110739848-001

Every chapter about the single TPE contains a part concerning the processing. That is important because the processing conditions have a significant influence on the mechanical properties of the resulting parts. Mainly, the processes of extrusion and injection molding will be discussed. They are the incumbent methods for TPE. Of course, the TPE enter into the trendsetting technologies of additive manufacturing like sintering or filament printing, to name but the most usual ones. On the other hand, TPE are expandable during the melt processing by the well-known methods, which is gas injection or adding of foaming agents. Even particle foams of different TPE are established in the market.

1.1 Rubber

In official communication, the term thermoplastic rubber is sometimes used, which is slightly confusing because a TPE is mentioned. Vulcanized rubber is always chemically cross-linked (Fig. 1.2) and can no longer be melted. The typical large volume rubbers like SBR (styrene–butadiene rubber), NR (natural rubber), or BR (butadiene rubber) have high molecular weight for technical application, which are barely achievable for TPE. During rubber processing to make rubber mixtures, the material will be degraded to bring in all the components and finally produce a vulcanized elastomer. After the cross-linking process, the chain length of the rubber is reduced which makes the material stiffer, depending on the density of cross-linking. Without this process, the rubber would flow under mechanical load.

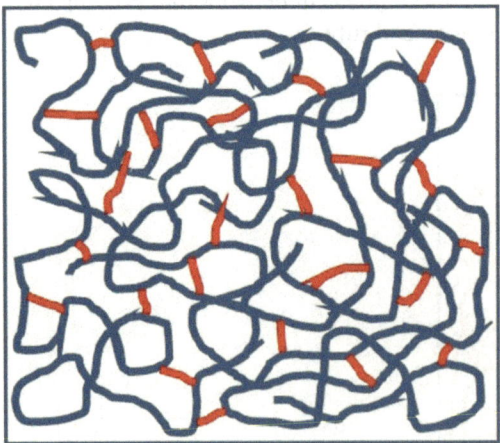

Fig. 1.2: Structure of a vulcanized rubber.

The recipe influences the hardness of such an elastomer and can range from soft, for elastic rods, to rigid, for golf ball cores. To maintain the elasticity while achieving an appropriate firmness, mineral fillers are used, for example, in car tires, active carbon black is added, which essentially acts as a semibound hard phase. Besides providing wear resistance, such a filler reduces the material degradation by UV light. All this has an influence on hardness and the elastic behavior of a made rubber. Achieving an easier handling or reduced material cost, inactive fillers are used as well, when the property request allows it.

NR and BR have a linear olefinic backbone and only a small degree of cross-linking is required to achieve a typical rubber elasticity which is characterized by high rebound forces and low deformation at high elongation. The molecular chains seek the lowest energy conformation, which is the statistical coil and it is called the entropic elasticity. Such materials closely follow Hooke's law – in which the relation between elongation and force is linear – which is visualized by an ideal spring. This behavior is connected to a good cold flexibility. NR or BR have a low freezing temperature around –100°C, typically referred to as the glass transition because the coils do not crystallize, rather they solidify like a glass. In general, elastomers should have a certain amount of elasticity below 0°C and should not break under mechanical load or upon impact above the glass transition.

Besides the soft elastic character of most of rubber materials, the covalent bond of cross-linking ensures that the polymer does not melt. This property provides a good dimensional stability at elevated temperature, which means a good heat distortion temperature. This is the primary difference from TPE. On the other hand, this also means that it is not possible to melt a vulcanized rubber for thermoplastic processing. This is a very important aspect in the non-neglecting environmental discussion, here the recycling. This is much easier for TPE, whether in the first step of reusing waste or non-conformant material in the plant already, which reduces the material cost as well. In the second step, used parts from the market can be reprocessed to make new TPE pieces. Another aspect is the cost advantage due to faster processing and easier handling of TPE pellets compared to rubber strips, the pre-product before transforming to final parts and the vulcanization step. Related to one of the biggest application field, the overmolding of a soft TPE on a rigid thermoplastic, they show these aspects very clearly.

This application (Fig. 1.3) sounds trivial, but nearly everybody is using it. No one can imagine not to have a soft grip on the toothbrush. The manufacturing in a multicomponent injection molding opens a good opportunity of a balanced cost–benefit balance and a lot of design possibilities.

Chapter 3 gives a deeper look into the difference of these both elastomer classes.

Fig. 1.3: Toothbrush from a rigid ABS overmolded with TPS of Kraiburg TPE (source: First Thai Brush).

1.2 PVC-P (PVC plasticized)

Polyvinylchloride (PVC) is a rigid thermoplastic material and needs a high amount of plasticizer to make it soft and elastic. It can be included into the family of TPE, but this polymer compound has existed for many years before TPE appeared, and PVC-P (PVC plasticized) remains as a material in the group of thermoplastics. Concerning cold flexibility and the elastic behavior, there is a little difference between PVC-P and TPE, but it is important to consider this material as it is used in similar applications as TPE.

When a user is looking for an alternative to PVC-P, TPE products are considered to be suitable. Of course, PVC is quite unbeatable regarding the price–performance ratio. It is easy to process when the temperature is not too high, the mechanical properties are often good enough, and its resistance against a variety of substances makes this material attractive. The environmental concern of PVC pollution, the sensitivity to corrosion during processing, and the possible migration of the plasticizer drive the desire to seek an alternative. Even the use of organic plasticizer is under discussion in the public. Furthermore, when PVC burns, dioxides are created, which are highly toxic. For all the aforementioned reasons, it is worthwhile to seek for and consider about alternatives.

1.3 Olefinic polymer compounds

The predominant polymer mixture of a TPO is a thermoplastic matrix of polypropylene (PP) and an unsaturated rubber like ethylene–propylene–diene copolymer (EPDM) or a saturated ethylene–propylene copolymer, made on a twin-screw extruder. The concentration of rubber is high, in order to provide a flexible material rather than an impact-modified thermoplastic. However, it is difficult to define a specific cutoff in these two regimes. In the case of more than 50% of rubber, EPDM would be the continuous phase and PP the disperse one. The soft TPO grades with a high rubber content (30–40%) are mainly used in extrusion applications where it is possible to process the melt under continual low shear. An example is the extrusion of films with an embossed surface to create a skin for instrument panels in automotive interiors. In a special procedure, the film is treated with an electronic beam to cross-link the material, improving the wear resistance of the surface. In injection molding, a change of morphology can take place during the processing and consequently result in a loss of mechanical properties. This can be attributed to the high shear under these circumstances. Therefore, only few applications in injection molding are visible in the market, for example, the bumper of a vehicle for a good shock absorption (see Fig. 1.4). Good elasticity is not the key property requirement for this application, rather the ability to take the energy of an impact (damping or shock absorption). In the same application, TPO

Fig. 1.4: Car bumper from olefinic compound with high energy absorption (source: monkeybusinessimages/iStock/Getty Images).

copolymers are used to modify PP. These TPO will be introduced separately in Chapter 4.

The softer the polymer mixture is, the larger the changes in morphology during processing are. In harsh processing conditions, the mobile rubber phase agglomerates, especially when the ratio is close to 50:50. The phases are in competition with one another to build the continuous one. To avoid this, the rubber particles can be chemically cross-linked during the compounding step. This is, in principle, the method to make a TPV, where the mobile soft phase is a high amount of little, discrete, and stable particles of a diameter from 1 to 10μm. During the process step in the PP melt, these elastic particles flow like a filler without any tendency of agglomeration and it is possible to include 40–80% EPDM into PP. The difference of the two morphology is illustrated in Fig. 1.5.

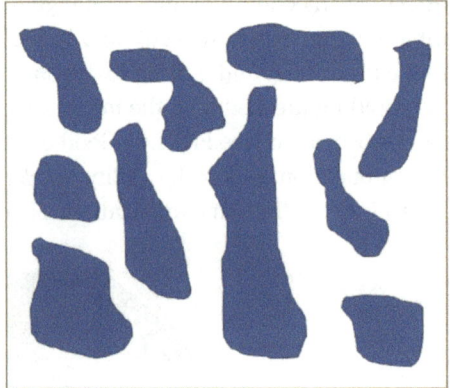

Fig. 1.5: Comparison of the two models of TPO blend (left) and TPV (right).

The share of rubber in the TPO compound could invite a discussion about TPE and impact-modified thermoplastics. A sharp distinction does not exist, but it is questionable how to classify the TPO compounds among the TPE family. Is the classification as a TPZ a better term because these are compounds? This will be an intensive discussion in the future and due to this uncertainty, TPO blends will not receive the same priority in the further course of this book compared to other TPE. The chapter on TPO will introduce the olefinic block copolymers and the suffix will hint as to how to modify the nomenclature of the different TPO in the ISO norm 18064. For the time being, the proposal in the Draft International Standard is to differentiate the TPO by the suffix TPO-M for the compounds and TPO-C for the copolymers.

1.4 Copolymers

The block copolymers among the TPE are built with a crystalline or amorphous hard segment and a flexible soft segment. In case of TPU, TPC, and TPA, the hard segment is often crystalline and is mainly responsible for the mechanical properties, primarily the modulus and the durability. The soft segment is made by oligomeric molecules (telechelics) with molecular weights of mainly 1,000–3,000g/mol and responsible for the flexible part of properties. Usually the chemical resistance is also influenced by the chemistry of the soft segment. TPO and TPS are constructed via polymerization of pure monomers and sequences of different blocks from monomers, in order to build hard or soft segments in the polymer chain, which are connected by covalent bonds. The nomenclature of TPE is related to the amorphous or crystalline hard phase for copolymers. In Fig. 1.6, the most well-known representatives of the groups are shown, whereas the chemical formulas are a typical selection of hard and soft segments.

Upon a deeper look into the norm EN ISO 18064, a more detailed fragmentation of TPE types is possible. For example, a TPS based on a polystyrene ("S") hard phase and a polybutadiene ("B") soft phase is named as TPS-SBS. A TPV containing EPDM as flexible and PP as the continuous hard phase is called TPV-(EPDM + PP). In Chapters 4 to 9, the members of the TPE family will be introduced more extensively.

It must be stated that this book does not serve as a comparison between the properties of different TPEs. Every member of a group has such a broad product portfolio that no comparison is comprehensive enough. This is only possible when a grade would be selected from different types of TPE for a specific application. A consultation with a dedicated supplier should be taken for every case. Here, the approach is taken to describe the character and behavior of selected TPE grades from every family in its own chapter.

Also, the huge area of additivation of TPE for achieving certain properties will not be touched. First, order the flame protection, then the antistatic or conductivity. Even when it must be taken into account that after incorporating such components the mechanical properties should be remained, such technologies and formulations open much more fields of portfolio in this product families. In this book, the relation of structure and properties is the key point.

Hereafter, there is an introduction of test methods that often appear in the book. A detailed description can be found in Chapter 11.

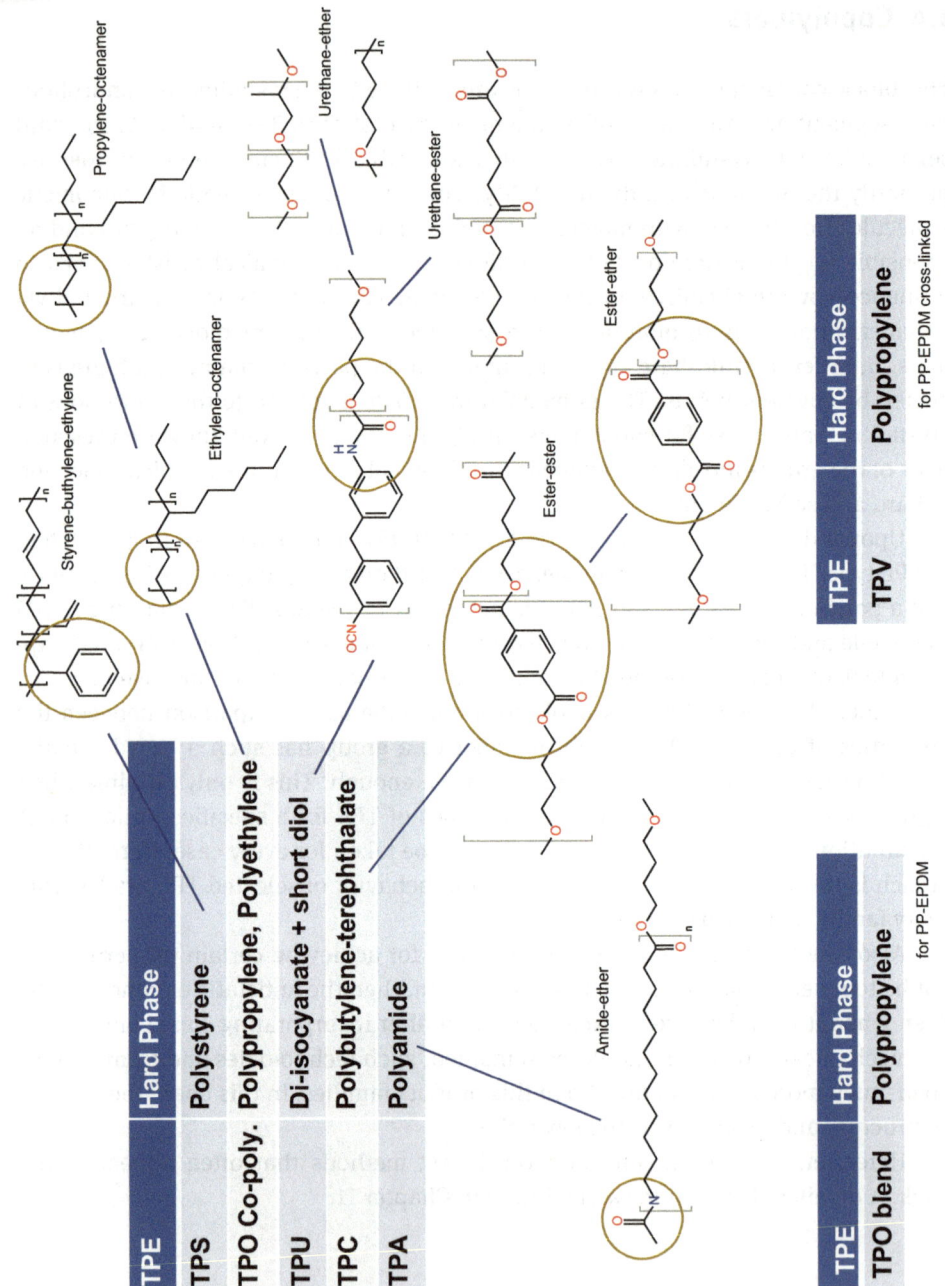

Fig. 1.6: Classification and chemical structures of TPE.

2 Characteristic methods

2.1 Shore hardness

The classification of elastomers is made usually by shore hardness, which is described in more detail in Section 11.1. This measurement by ISO 48-4, formerly ISO 7619-1, selects the materials by the level of hardness. It is a hardness of the surface and not a true physical unit, rather a classification and a helpful orientation, because the measurement is very easy, quickly performed, and the values are well accepted in the industry. A test bar of sufficient thickness is placed under load with a needle (pin) and a defined weight for 3s or 15s. For harder materials of a Shore D, a sharp pinpoint is used. When Shore A or softer materials are measured, a flat pin is used. It should be mentioned that there is no mathematical relation between these two hardnesses, but it is recommended at high Shore A (≥95A) to provide Shore D as well.

Normally, TPE materials are available in a range of Shore 30A to Shore 80D. Highly plasticized TPS compounds have even lower hardness (Shore 0) and are almost gels. Such samples are classified in Shore A0 (or Shore 00), where a rounded probe is used to test shore hardness. There are few more shore ranges possible but these are rarely used in the TPE business.

The comparable norm for plastics and hard rubber articles is the DIN EN ISO 868.

2.2 Tensile strength

The general norm regulation to measure the tensile properties on plastic materials is ISO 527-1, where the graph of a hard, tough, and elastic polymer is illustrated as an example (see Fig. 2.1). Opposite to hard plastics, the tensile–elongation curve of an elastomer has a completely different character. The tensile behavior of a stiff material has a steep upward line, achieves a maximum, and decreases slightly before the test bar breaks. The considerably linear increase at the beginning makes the detection of elastic modulus possible because the linear elastic area (Hookean range) is long enough. This is different for elastomers, especially in case of soft products. At the beginning of the tensile measurement, the force changes quite early at small elongation until a near-plateau state is reached. Then, in case of many TPE grades, the tensile force increases incrementally until the bar breaks. Just before that happens, the tensile strength will be detected in megapascal and the elongation at break in percentage (%).

The shape of the test specimen is usually Type 5A (ISO 37: Type 2) or S2 (DIN 53504), and the elongation is detected at the smallest dimensional area. The comparable norm for rubber articles is ISO 37.

https://doi.org/10.1515/9783110739848-002

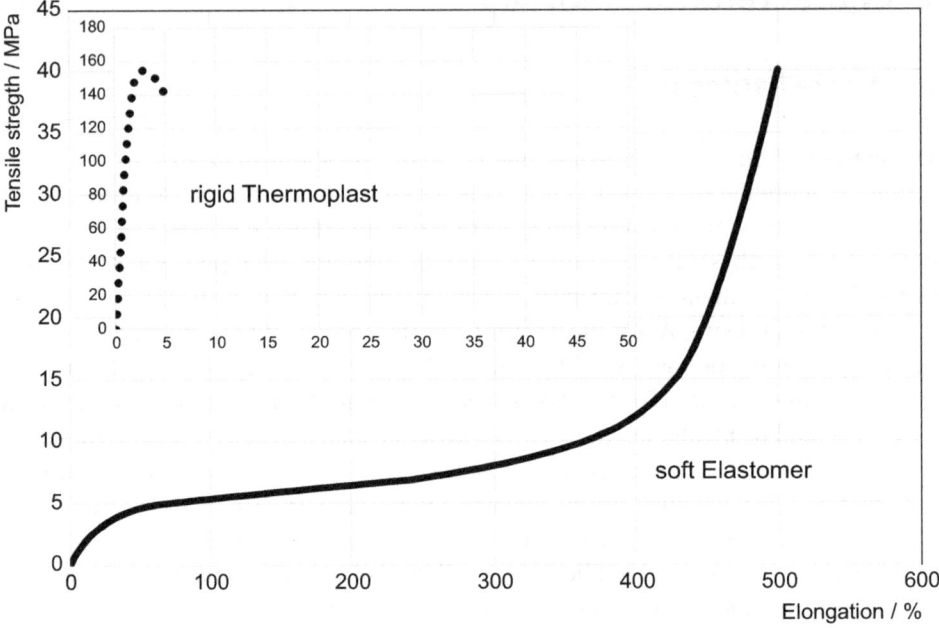

Fig. 2.1: Comparison of different stress–strain characteristic of elastomers to rigid thermoplastics.

These characteristic curves will not be presented in the individual TPE chapters, rather than the typical value ranges to show the broad range of properties. Various tensile–elongation curves will be shown intermittently throughout the book.

2.3 Intermittent stress–strain measurement

A nicely graphical method to describe the elastic properties is the intermittent stress–strain measurement. This graph shows the level of elasticity compared to the applied elongation, more precisely, the increase of the viscous portion together with the increased elongation in a kind of hysteresis. Also, this method is helpful for the user working with elastomers, seeking characterization in advance.

In the present model (Fig. 2.2), a specimen will be repeatedly stretched under increasing elongation in tensile mode in universal testing equipment with subsequent relaxation. For an ideal elastomer, the tensile value would return to the original point after every step, but all plastics are nonideal and deform under load, even TPE and rubbers as written above.

The residual elongation will be plotted over the applied elongation to track the deformation behavior of the sample. Clearly, the curve in Fig. 2.3 is flatter, the higher the elastic content of a specimen is. Sometimes a break in the shape of the curve exists, exhibiting two regimes, as depicted in that figure of a TPU. This point can be

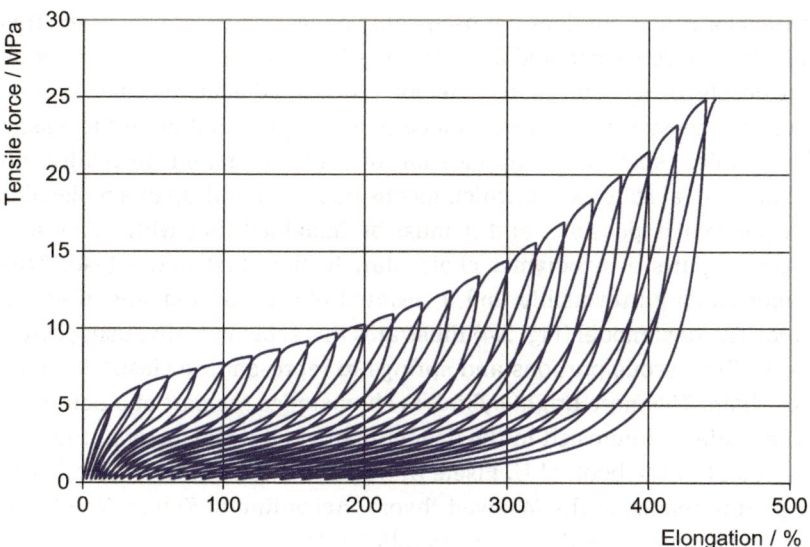

Fig. 2.2: Intermittent stress – strain measurement of a TPU (Shore 90A – 3s).

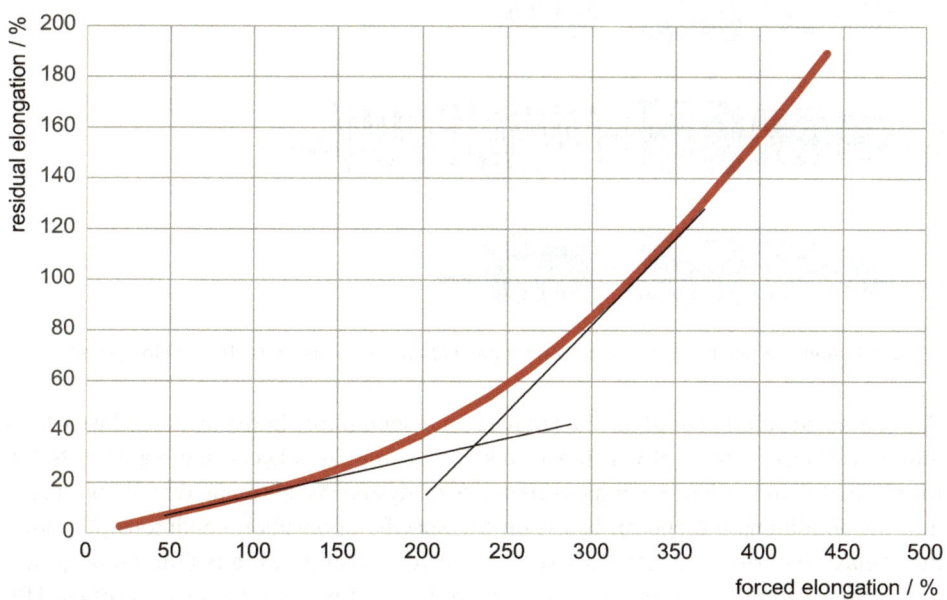

Fig. 2.3: Plotted strain ratio of applied strain versus residual strain from the intermittent stress–strain measurement of a TPU (Shore 90A – 3s).

called the critical elongation but does not necessarily appear for every elastomer. This point can sometimes be very clear, and at other times it is hardly visible. It will not be discussed more deeply. In Section 11.4, this method is described in more detail.

In general, this measurement creates a meaningful representation of the elastic behavior of a sample, especially in application of high elongation. In reality, no ideal material in an ideal state exists, which means no ideal build-up of a molecular structure of a polymer is possible, and it must be imagined that with increasing strain, the weakest parts of a polymer chain start to flow first under load. This changes the morphology irreversibly and a residual elongation remains after the load is removed. The next model (Fig. 2.4) illustrates this schematically, using cylinders to represent the viscous portions and springs to represent the elastic sections of the polymer chain. The more bright cylinders characterize the part of a deformed structure. If the reader is interested in the deeper mathematic phenomena, the fundamentals are found in the book of U. Eisele (*Introduction to Polymer Physics*). The model in the picture relates to the Maxwell theory. According to Kelvin–Voigt, the viscous and elastic elements are drawn in a parallel order.

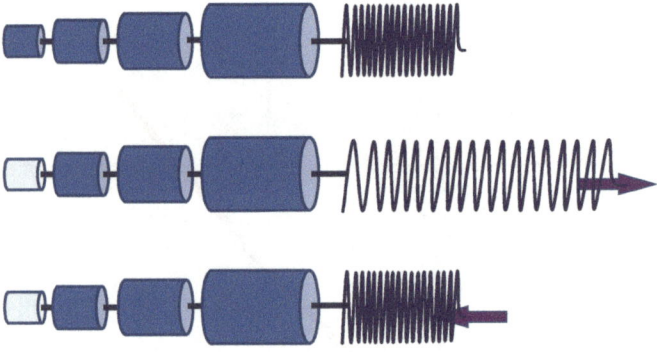

Fig. 2.4: Model of a plastic deformation (bright part) of an elastomer under forced elongation.

The characterization of elasticity has played an essential role in the textile industry for a long time. Elastic fibers, of which a well-known example is Lycra, are regularly tested for residual deformation. Such measurements of hysteresis are made directly on fibers, and the conditions are chosen based on the specific application conditions. Measurement may be performed statically or dynamically. Usually, the residual elongation of the fiber sample is observed after repeated stretching to 50% several times. TPE fibers directly spun from an extruder are quite similar and are used for fabrics as well (Fig. 2.5). The elastic behavior is obviously an important aspect.

Lycra is produced from a polymer solution in an organic solvent, precisely it is a polyurea. TPE alternatives employ a melt spinning process directly from an extruder and produce, therefore, a clear environmental advantage, bypassing the use of organic solvents. This has generated increasing interest in the market.

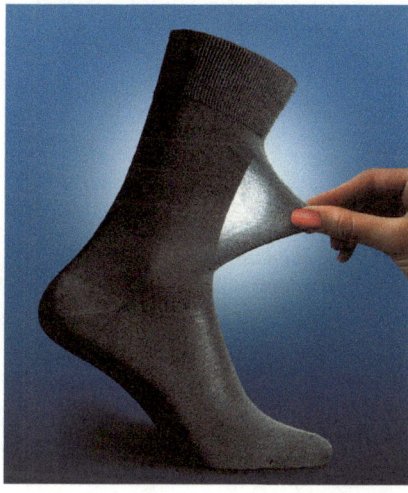

Fig. 2.5: Elastic sock made from melt spun elastan fiber (source: BASF).

2.4 Dynamic mechanical (thermal) analysis

One essential element to principally characterize representative TPE materials is to consult the dynamic mechanical analysis (DMA), which is a method that analyzes the dynamic strength of a sample over a wide range of temperatures. It starts where the material is in a glassy state at low negative temperature, transitions into an elastic phase, and ends upon softening at high temperatures. From every product group, a few typical TPE are drawn to describe the relation of structure and property via DMA. All measurements on the same equipment relate to ISO 6721 under the same condition in a frequency of 1Hz. The temperature is increased in steps of 5°C for every subsequent point. This provides a conditioned specimen at each point of evaluation. The more detailed description of that method is in Section 11.3.

The next model curve (Fig. 2.6) shows DMA of a polyether-based aromatic TPU as a representative illustration, where the different phase transitions are shown. The storage modulus (see Section 11.3) is plotted in a logarithmic scale because of the deep difference before and after the glass transition to illustrate it in a clearer arranged kind. Beginning at low temperatures, the coil of frozen soft segments, an amorphous glass, moves from a brittle inflexible state into an elastic one. This is called the glass transition of the elastic part. In the following regime, the soft phase is responsible for the elastic behavior and the hard segments are still crystallized or frozen keeping the material in a solid elastic state. This temperature regime may be referred to as the "working" or "application" temperature. When higher temperature achieves the melting point of the hard phase, the material starts to soften or to melt and cannot be mechanically measured further. At this temperature, the product cannot be used any further and enters the phase for melt processing. It is recommended

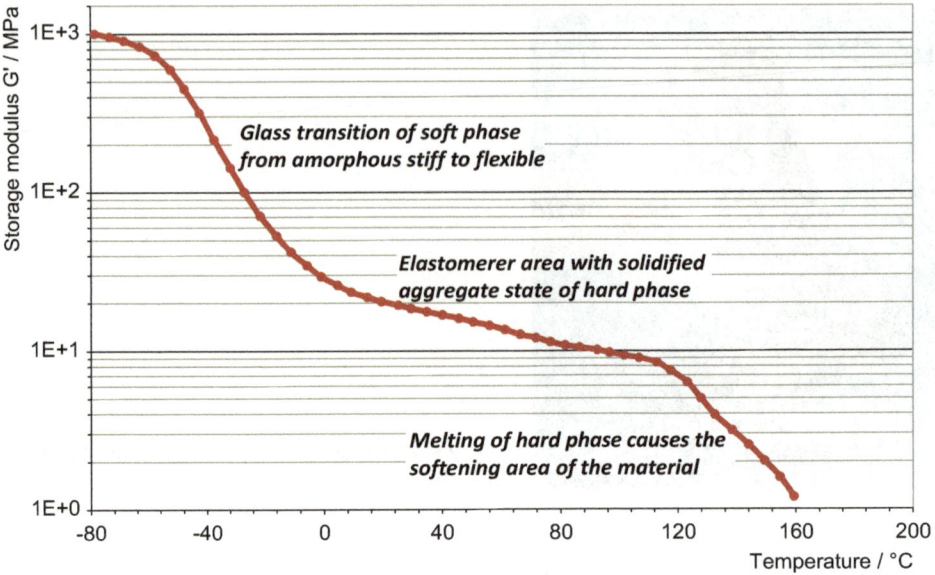

Fig. 2.6: DMA of a TPU (Shore 90A – 3s), measured in torsion mode on a test bar with a frequency of 1Hz.

to first estimate the ceiling temperature of a sample before conducting a DMA measurement. A thermomechanical analysis (TMA) is best suited for this.

It was mentioned in the prologue that the best illustration of elasticity is a steel spring and is described by Hooke's law under ideal conditions. There, the elongation is linearly proportional to the applied force. For elastic polymers, the same correlation should be valid but the load leads to disorientation, to some extent, and a change in the material morphology. This behavior is a kind of shift in orientation of the matrix, which means it flows. The scientific term is damping because the applied energy would not be returned completely. For this reason, TPE are always viscoelastic materials. It can be seen in a DMA analysis, in response to the applied dynamic torsion because a share of energy dissipates in the matrix, which is the viscous part. The phase deviation is called the loss factor, or tan delta (tan δ) which is the ratio of the viscous and elastic modulus. Above a value of 1, the material leaves the solid state and transitions to a liquid. Understanding this method and the results is very helpful for material selection and the development of technical solutions in the desired application.

A typical discussion of DMA on TPE would not consider the effect of measurement at higher frequencies but what about this influence? It is known that the glass transition moves to higher temperatures at higher frequencies in DMA measurement. Use of the time–temperature–superposition principle allows for data extrapolation outside the possible frame of standard measurement conditions. The shift factors are incorporated in the equation of Williams, Landel, and Ferry and apply well to homogenous polymers.

Figure 2.7 illustrates the DMA of a TPU Shore hardness 85A, which is measured at 0.1, 1 and 10Hz. The curve is reduced to the area of the glass transition and it shows an increase of the point of inflection of about 5°C on every step. That means the higher the frequency impact on a TPE is, the worse the cold flexibility comes out.

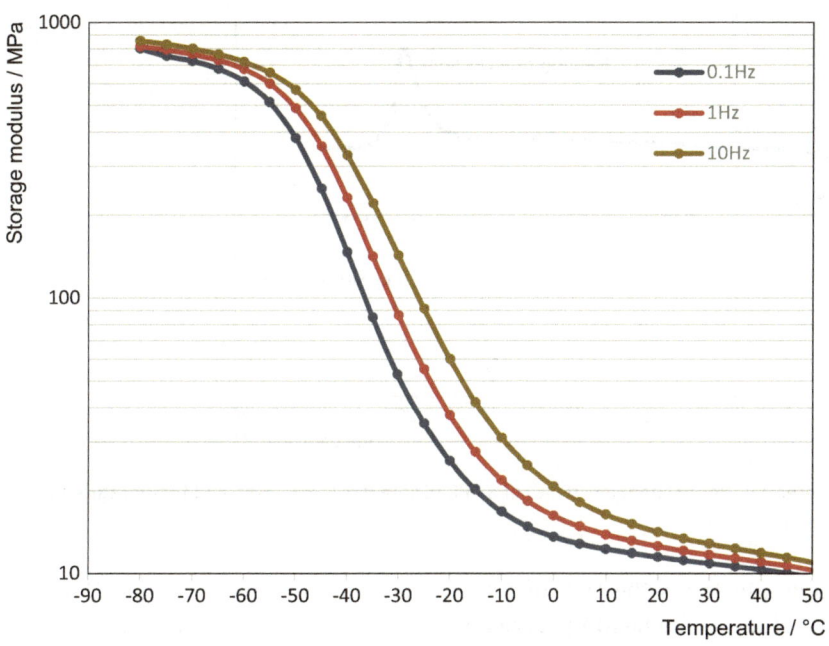

Fig. 2.7: Dynamic mechanical analysis of a TPU (Shore 90A – 3s), measured in torsion mode on a test bar with a frequency of 0.1Hz, 1Hz and 10Hz, area of glass transition.

2.5 Differential scanning calorimetry

For some product families of TPE, it is worthwhile to learn more about the thermal behavior before processing the pellets. A suitable thermal analysis is described in the norm DIN EN ISO 11357-1 at a basic level. More precisely, it is differential scanning calorimetry (DSC), in which the polymer sample and a reference or blank, are heated simultaneously (see Section 11.5). The reference sample should ideally not have any thermal effects in the desired temperature range. Thus, the heat transitions of the polymer sample are detectable. In case of an elastomer, a glass transition appears at low temperature, when the frozen polymer coil softens, and another transition at high temperature, where the melting of the hard segments are visible. After the first heating has been completed, the first cooling is conducted and the main transition that appears is the solidification of the hard phase. Usually, the heating and cooling runs are performed at a rate of 20K/min. The closer the melting and the solidifying ranges are to each other, the earlier the polymer melt becomes solid after processing, indicating that a shorter cycle time should be expected.

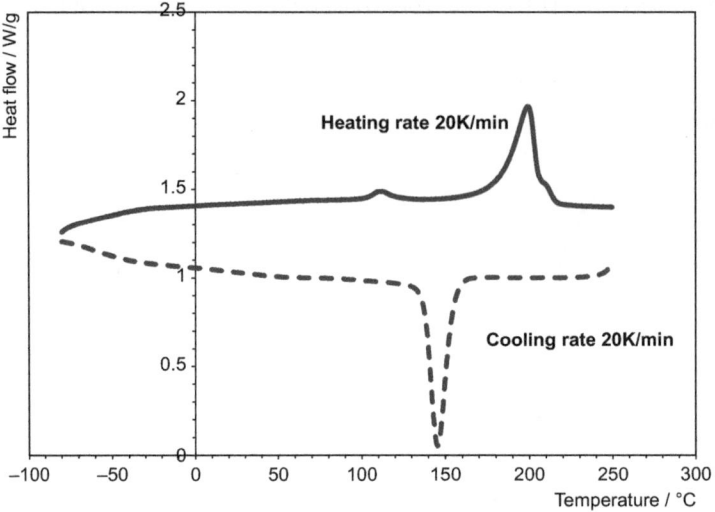

Fig. 2.8: Differential thermal analysis (DSC) of a TPC (Shore 97A/48D – 3s), and temperature ramp rate of 20K/min.

This example illustrates the DSC of a thermoplastic polyester elastomer (Fig. 2.8) where the melting in the heating curve and the distinct recrystallization in the cooling curve is clearly to identify.

Here, it is mentioned that the DSC will be drawn for TPU in few cases only because the morphology of the polymer changes during the measurement and a suitable analysis can only be done partially.

3 A principal comparison of TPE to elastomers (vulcanized)

In the prologue, a brief fundamental difference between rubber and thermoplastic elastomers (TPEs) have been discussed. In both cases, we talk about elastic material, but a finished rubber is no longer thermoplastically processable. A differentiation must be made between nonvulcanized (caoutchouc) and the vulcanized one, which is utilized in technical applications. On a molecular scale, the caoutchouc is a structure of statistical coils. It is not practically useable because under mechanical load, these molecular chains slide apart. To enable practical use, they are chemically cross-linked before applied in technical function as a rubber article. This process, called vulcanization, in an industrial scale takes place with sulfur, peroxides, phenol, and in some cases phenol formaldehyde as well. The last one plays a role in the TPV (thermoplastic vulcanizate) technology, besides peroxides (see Chapter 5). These cross-linkers will be mixed into the pure rubber together with some additives like fillers, oil, stabilizers, accelerators, and processing aids at low temperatures in an inline mixer (kneader) or on a roll mill. The resulting sheets or strips will be sent to the processing factory where an injection molding machine or an extruder will be fed, and the sheets are carefully formed into the desired shape. Subsequently, the resulting pieces must be heated to start the vulcanization process. From the extruder, the extrudate continuously moves through a hot bath. In the instance of injection molding, the mold is heated. Additional important molding processes exist, like the hot press for big rubber parts, well known for car tires. Electronic beam vulcanization is seldom done on an industrial scale. The complexity of these steps makes the advantage of melt processing a TPE clear, namely, easier handling of pellets and a faster cooling time, as opposed to a long vulcanization.

Besides fillers and plasticizers, the level of cross-linking determines the hardness of the product. This allows the same material to be used for a softer rubber band of low cross-link density, and a hard, highly cross-linked golf ball core. The long main chains of a rubber molecule with a high mobility create a good elasticity which decreases with increasing number of transverse bonds, that is, cross-links. Thus, a rubber is never brittle, independent from its hardness because the backbone remains flexible. The high level of rubbery character is called entropy–elasticity, where the polymer tries to achieve the lowest energy conformation after every relaxation, in other words, the statistical coil. This is similar to the soft phase of a TPE and also permits entanglement, depending on the chemical structure. Compared to soft rubber articles, the flexible chains of a TPE are shorter and the elasticity less pronounced. Looking at the intermittent tensile–elongation curve in the next diagram (Fig. 3.1), an even softer thermoplastic urethane (TPU) of hardness 60A shows a higher creep than a soft rubber mixture out of natural rubber and polybutadiene rubber (NR:BR = 3:2, 29.4% carbon black, 4.1% oil) of hardness 65A.

https://doi.org/10.1515/9783110739848-003

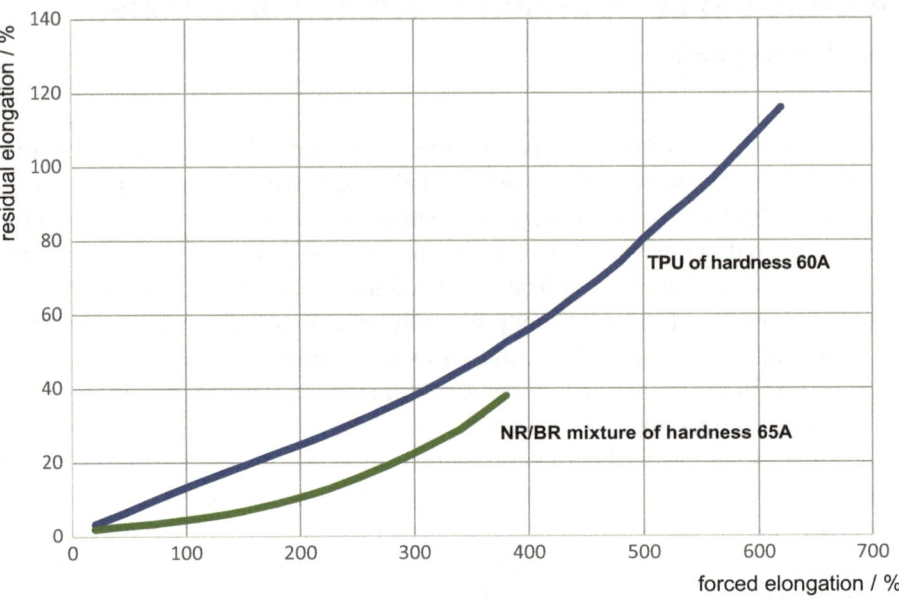

Fig. 3.1: Comparison of forced and residual strain relations from the intermittent stress–strain measurements of a vulcanized NR–BR–rubber (Shore 65A – 3s) and a TPU (Shore 60A – 3s).

Though the mixed rubber sample is a harder material, the slope of the curve at the beginning is much flatter. NR and BR are very high elastic caoutchouc types with a very low viscous component under forced elongation. Until an elongation of 100 %, there is hardly any deformation, and a low degree of residual elongation is detected. This is the consequence of the long linear polymer chains of the rubber in comparison to the soft TPU. Additionally, the presence of a chemical bond in the rubber inhibits creep more effectively than the physical cross-links of the semicrystalline hard phase of a TPU.

Nevertheless, rubber also exhibits creep and this occurs due to the interaction with the large amount of filler in the rubber. The viscous component increases under higher stretch and the rubber softens. Carbon black particles move and remain in the new position in the polymer matrix after relaxation. This is called the Mullins effect, which is increasingly evident with higher filler content. Unfilled vulcanized NR displays close to an ideal elastic behavior with nearly 100% resilience. This behavior is combined with a constant modulus over a wide range of elongation. This property is advantageous for gymnastic equipment and resistance bands for physiotherapy.

Dynamic mechanical analysis (DMA) was previously introduced as a good tool to characterize an elastic polymer. Therefore, the same materials, the NR–BR–rubber and the TPU, are compared in Fig. 3.2. A quite similar modulus can be seen.

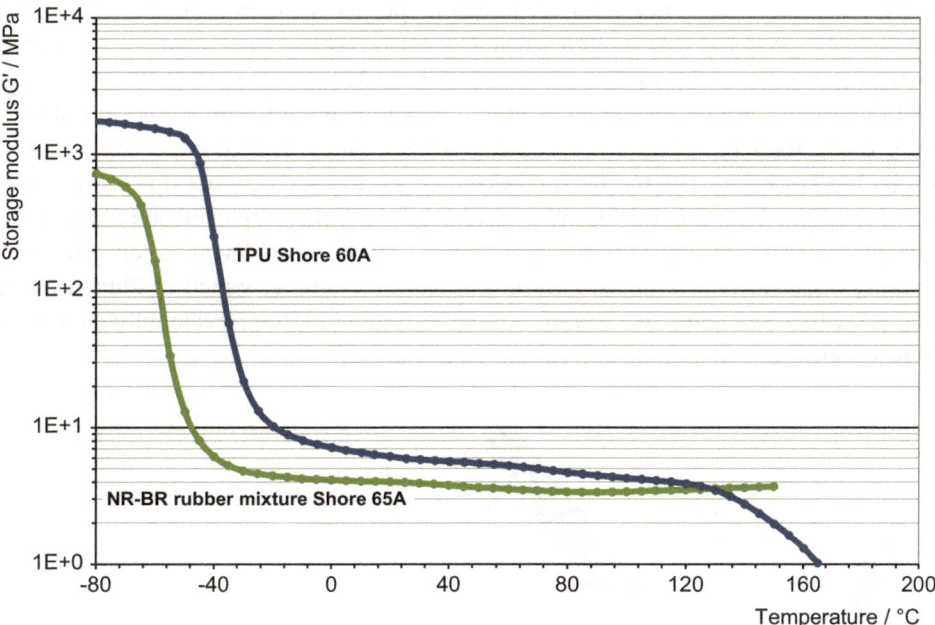

Fig. 3.2: Comparison of the DMA of a vulcanized NR–BR–rubber (Shore 65A – 3s) and a TPU (Shore 60A – 3s) on a test specimen in torsion mode at a frequency of 1Hz.

Both elastomers show a reasonably constant modulus over a wide range of temperature with slightly better character for the rubber sample. In the high temperature range of the curve at 120°C, the thermoplastic material enters the melting phase and the NR/BR maintains its modulus until the degradation or distortion of the polymer chain. On the left side, the glass transition is illustrated, and significantly better cold flexibility of the rubber is depicted by its lower temperature transition, even for materials with similar modulus.

For increased strength and better handling, a high amount of carbon black is incorporated into the rubber mixture. The filler is very porous and has a large surface area to bond to the polymer molecules as much as possible, when distributed homogeneously. Normally this is innate for the TPE because the hard phase acts as a nanoscale filler and is bound to the elastic soft phase in case of copolymers TPS (styrene block copolymer), TPO (olefinic block copolymer), TPU (urethane block copolymer), TPC (ester block copolymer), TPA (amide block copolymer). This is another story for TPE compounds (TPO and TPV) where the morphology is completely different, which means the elastic part is done by an added elastomer. Of course, the bonded nanofiller of the TPE copolymers can melt at elevated temperatures depending on the hard phase structure. Additionally, the chains of the semicrystalline hard phases may slip into an alternate configuration under a load, depending on the temperature. Creep of the polymer occurs unlike a well-vulcanized rubber, which can be used for highly loaded seals, whereas TPE are not suitable. Ethylene–propylene–diene rubber (EPDM), for example, is a material that is often seen as an elastic sealant material at elevated temperature. A good compression set of lower than 10% at above 100°C is possible and in case of high performance elastomers like nitrile or fluoro rubber resistance until 200°C are achievable. A TPE could be an alternative only at lower temperature. Under less severe conditions, TPE materials are advantageous to use due to their easier processing, including the ability to overmold a stiff thermoplastic body in a direct injection molding or extrusion process. This is useful for achieving a soft grip for a handle, a gasket on a cap, or a sealing edge on a profile, which is illustrated in Fig. 3.3 with a TPS edge on a EPDM sealing. Furthermore, the recyclability continues to be a significant advantage and cannot be neglected.

Fig. 3.3: Overmolding on an edge (blue) of an EPDM car window profile with a TPS-M (source: Allod).

The biggest use of rubber is in tires for any vehicle, mainly cars, trucks, and aircraft, even bicycles and motorcycles, not to forget belts for conveying and transportation. On the one hand, a tire's rapid velocity absorbs such considerable energy that the increasing temperature would start to soften the TPE. On the other hand, the static impact of a load in a warm climate can initiate an undesired creep and deformation of the wheel. When one considers the wide temperature range and versatile environmental wet and dry conditions, a tire tread is required to withstand, only a rubber compound is suitable. For a long time there have been attempts in the market to replace the rubber of a car tire with thermoplastic materials, capitalizing on the faster manufacturing and the better reuse after the lifetime of the tire. One example is "Tweel" from Michelin, who has spent a considerable amount of time developing this concept and has only recently revealed a prototype. One can imagine the big hurdle it is to make such a tire by an injection molding process.

The special construction of the spokes in Fig. 3.4 obtains sufficient cooling from the air flow while in operation, but the tendency of creep can become a challenge while parked in the sun. As mentioned, some prototypes are poised to compete against the classical pneumatic rubber tire, but it will still take some time and development effort to achieve a successful replacement. The car manufacturer, GM, announced the launch of a successor of the "Tweel", the "Uptis," in few years on its cars. The question will still be whether the spokes can be made of a TPE rather than of rubber or a cast elastomer. The tread, at any rate, will stay with a suitable rubber formulation.

Fig. 3.4: Car tire "Uptis" (source: Michelin).

In general, TPE suppliers raise a subtle claim to substitute rubber materials in many technical applications by emphasizing the advantage of easier processing. This is the reason that there is still a colloquial term "thermoplastic rubber" when TPO blends, TPV, or to some extent TPS compounds are mentioned. The olefinic character of these TPE justifies this terminology. Therefore, the formulations of these rubber-like grades are also established with the same weight units, "phr." Historically this originates from the rubber industry and means "parts per hundred rubber." Here, the sum of recipe parts does not end up at 100%.

4 TPO – olefin-based TPE is introduced

In Chapter 1, it was mentioned with regard to the thermoplastic elastomer (TPE) structure that the olefinic compounds (TPO-M) will be minimally discussed, and more emphasis will be placed on the TPO copolymers (TPO-C). Along with styrene block copolymers, both TPO and their copolymers are the largest group among the TPE family (see Chapter 12).

TPO copolymers are a fast-growing family because they can be produced with cost-saving advantages directly from the cracker from their olefinic monomers, which are α-olefins like ethylene, propylene, or 1-butene, as well as higher homologues like cyclooctene or norbornene. From there, block-copolymers can be built up to various macromolecules with Ziegler–Natta or especially metallocene catalysts. Both the semi-crystalline and the flexible phases can be made from polypropylene (PP) or polyethylene (PE) blocks. Determination of the composition depends on the molecular weight of the segment or its stereospecific order. For example, if the PP block is isotactic, it will be the more crystalline part, as illustrated in Fig. 4.1 about Vistamaxx from ExxonMobil, whereas the character of the PE segment is driven by its molecular weight, whether it creates the hard or the soft phase. This counts when no other elements are incorporated into the chain.

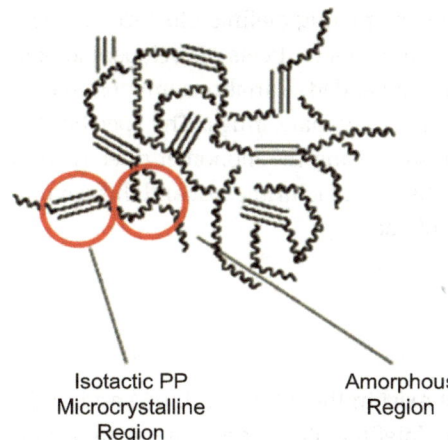

Isotactic PP
Microcrystalline
Region

Amorphous
Region

Fig. 4.1: Structure of TPO-C (copolymer) with crystalline PP and flexible PE segments (source: EP 0347128 B1 – ExxonMobil).

https://doi.org/10.1515/9783110739848-004

4.1 Manufacturing process

The decision to make the material by a batch process in a vessel or by a continuous one in a pipe reactor depends on the desired product's properties and the needed flexibility. The reaction occurs in a solvent or directly in mass production. The raw materials are in a liquid state and can be stirred. The resulting quality and polymer structure are determined by the process parameters and the catalyst, or a second one, used.

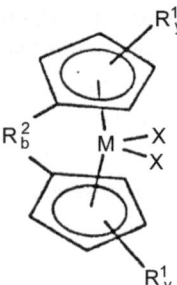

Fig. 4.2: Structure of metallocene catalyst (source: EP 0347128 B1 – ExxonMobil).

The Ziegler–Natta catalysts (e.g., titanium chloride with lithium–alumina–alkyl) are mainly suitable for a single polymerization initially propagated from the starting catalyst. The metallocene ones are more flexible in constructing olefinic block-copolymers (OBC). These substances, pictured here (Fig. 4.2), are mostly sheets of cyclopentadienyl layers that have a metal atom (often zirconium) embedded. Variations are possible by substitution of different ligands on the metal or the aromatic rings. The specific efficiency can be controlled by such kind of chemistry. Using the monomer directly from the oil cracker opens the opportunity to reduce the manufacturing cost and this circumstance is the basis to call these TPO a "reactor polymer."

4.2 Properties

The nomenclature of TPO copolymers does not exist in the ISO norm 18064 and in the industry, rather polyolefinic elastomers or OBC (olefinic block copolymers) are generally used terms, although a draft version in ISO is published to use the expression TPO-C for this TPE family (see also cap. 12 Epilogue). The differentiation between statistical and block structures does not appear in every case. The majority of grades offered in the market are mainly ethylene, propylene, 1-butene, and 1-octene. In all cases, they are nonpolar polymers that are resistant to polar media, such as water-based liquids. In contrast, they swell or dissolve in fuel or oils. Their resistance to degradation by light or UV irradiation and environmental conditions is similar to polyolefins in general. This opens the door for application in packaging or in the automotive interior where a moist atmosphere is expected. The high transparency, the flexibility, and the polar structure can be used for food contact in many cases (see Fig. 4.3).

Fig. 4.3: Fast food container of TPO (source: Scholz).

When considering the mechanical properties, the molecular weight, hard phase structure, and content play a significant role. Subsequently to the customer request, it is a combination of desired properties and good processing. The easier the thermoplastic processing and the lower the viscosity in the melt are, the weaker the mechanical strength can be expected. The task of the supplier is to find the best balance.

In a previous chapter (Section 2.4) the dynamic mechanical analysis (DMA) is presented and that it provides a picture how the material behaves over a wide range of temperature. Phase transitions can be seen and the level of modulus under low stress as well. In Fig. 4.4, the DMA of two TPO copolymers of different hardnesses is illustrated (for explanation of that method, see Section 11.3).

The softer material in Fig. 4.4 shows a clear glass transition and later a short elastic plateau compared to a rubber mixture of NR/BR. The harder sample changes in modulus value continuously. With this material, a broadly amorphous structure is expected; therefore, lack of distinct phase transitions with increasing temperature is reasonable because not a marked phase separation exists. This is also a criterion to have a material easy to process. At the end of the curve, the solid phase starts to soften, marking the highest possible service temperature. The DMA also discloses that these TPO are flexible below 0°C and will not become brittle in a frozen environment.

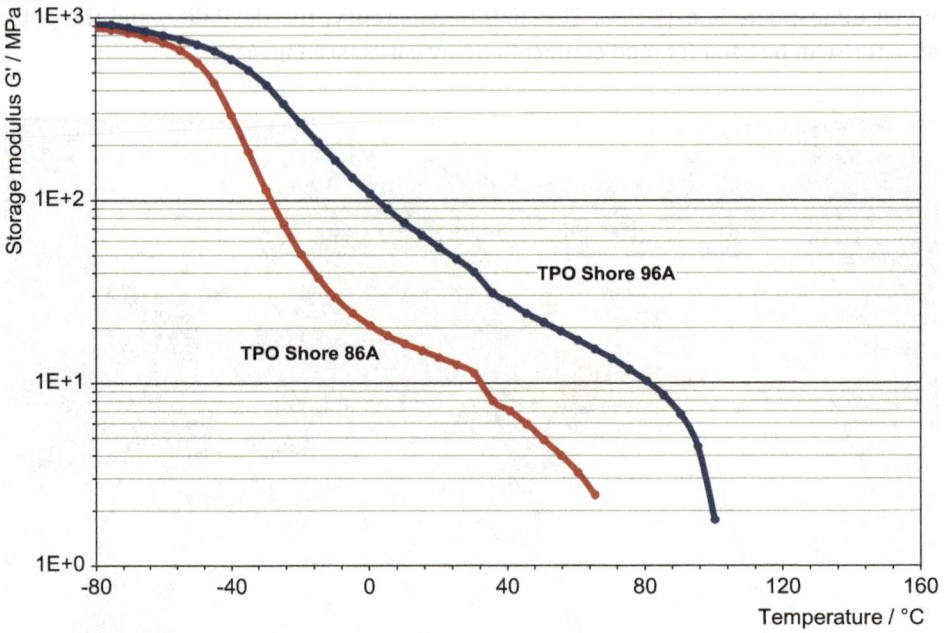

Fig. 4.4: Dynamic mechanical analysis (DMA) of ethylene–octenamer TPO polymers from Dow Chemical Co. (Shore 86A – 3s) and (Shore 96A – 3s), and test specimen in torsion mode at a frequency of 1Hz.

Even when technical applications are limited for such products, they are often used as impact modifiers for thermoplastics to make them ductile. With such character of easy melting it helps very well to mix an elastic component into a rigid polyolefin and to help them getting less brittle. To overcome this mentioned property limitation for the use as a single material, it is recommended that TPO are mixed with fillers or other olefinic thermoplastics to enhance the mechanical strength and to improve the processing behavior. Usually PP is selected as thermoplastic, and calcium carbonate or talc are chosen as inorganic additives. Pure TPO has a density as low as $0.86g/cm^3$, and as to be expected, an inorganic filler increases the overall density.

Mechanical strengths measured from tensile testing and compression set are typical characteristics to evaluate the utility of a material. Taking values from tensile strength testing (DIN 53504, ISO 37, ASTM D 412), the whole product family for a hardness range from Shore 50A to Shore 90A as follows:

Tensile strength	2–25MPa
Elongation at break	500–2,000%

Compression set values are not available for this material, as these are not a relevant characteristic in TPO applications. In this test, continual compression of test specimens at elevated temperatures will indicate their performance in a load-bearing application and is qualified by the change in thickness before and after this compression (see Section 11.2). This is not the big strength of a TPO. Determination of the material's yield behavior under compressive load is not well described by a DMA measurement. It merely hints at the expected transitions in a given temperature range.

Furthermore, the elastic properties of TPO are interesting. In Chapter 2.3, the intermittent stress–strain measurement was introduced as a useful method to characterize the elasticity of a material. In the following curve (Fig. 4.5), the level of residual deformation after stretching the TPO-C samples can be visualized (for explanation of that method, see Section 11.4).

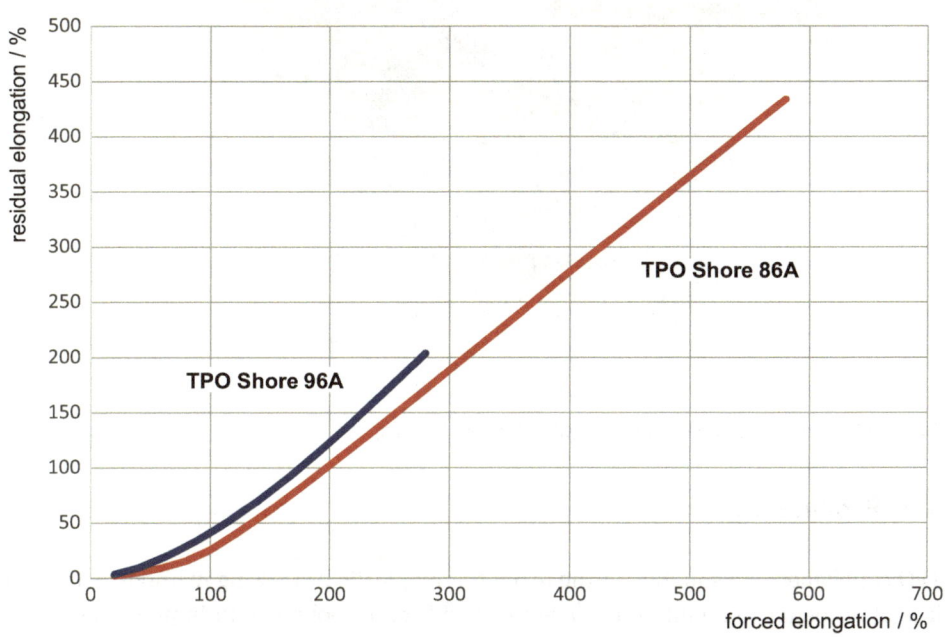

Fig. 4.5: Strain values from the intermittent stress–strain measurement of TPO-C from Dow Chemical Co. (Shore 86A – 3s) and (Shore 96A – 3s).

At the beginning of the cycle at small elongation, the graph remains flat and exhibits nominal elastic behavior. Thereafter, the morphology is continually deformed as the specimen is extended. The harder the product is, the more this effect occurs, as a consequence of the smaller content of flexible soft phase. Clearly, filler content also influences the elasticity. In general, these TPO have a perceptible level of elasticity at low and increased deformation at higher elongation.

In predominant applications of TPO, a high elasticity is not required. Higher priority is placed on the soft feel of an overmolded article and its sustained adhesion to the stiff substructure. The ability to high residual deformation can be used for stretch film in packaging, as illustrated in Fig. 4.6, where a good adhesion and a low permeation are required.

Fig. 4.6: Stretch film of TPO (source: Basilios1/Getty Images).

4.3 Processing

TPO copolymers are suitable for the usual thermoplastic processing methods, although it is often better to modify these materials with other polymers to improve their processability and adaptability to meet the user-defined properties. The manufacturer may adjust its processing parameters in advance by learning the thermal behavior of the material. It is helpful to know the melting and the freezing ranges of the sample. Such information is identifiable from a differential scanning calorimetry (DSC) measurement, described in Section 11.5. The heating curve in the following diagram (Fig. 4.7) presents a quite sharp melting peak with a wider range of premelting of the sample, which indicates a broad crystal size distribution. The recrystallization peak in the cooling curve is close to the temperature range of the melting of that sample. This indicates a fast solidification and the long way back to the baseline shows a slow recrystallization time. Fillers and nucleating agents help to reduce such effect but the low melting temperature is the indicator for a good processing under low energy.

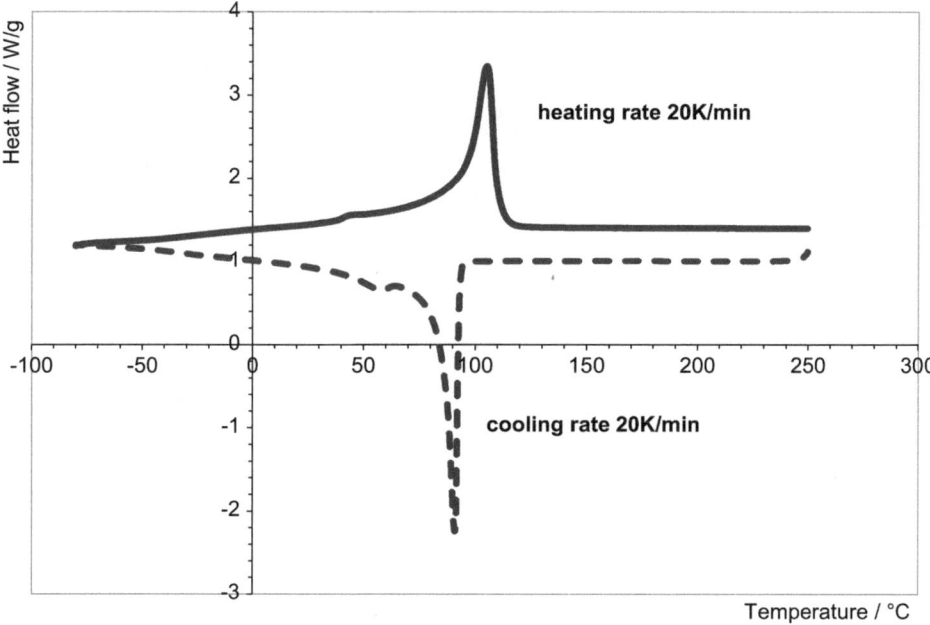

Fig. 4.7: Differential scanning calorimetry (DSC) of TPO-C from Dow Chemical Co. (Shore 96A – 3s), at a temperature ramp rate of 20K/min.

Thus, the thermal conditions can be readily adapted to the polymer material and the TPO will remain stable during the processing step. This results in good mechanical properties, which are published by the vendor's data sheets. Under harsh conditions, the TPE can be degraded. In addition to the values from DSC, the viscosity of the melted polymer is very important. The most widely accepted value is the melt flow rate (MFR) or melt flow index, and in Section 11.6 this method is described in detail. The MFR is a value of viscosity at only one point of shear and often not very helpful for a complex processing step. In that case, it is recommended to get the whole curve, where the viscosity is plotted in a graph as a function of the shear rate. This gives much more information about the processing behavior. With enough experience with the specific TPE, the MFR may be sufficient because the operator expects a good homogeneity and a lot-to-lot consistency. It takes much less effort to take an MFR measurement than construct the whole viscosity curve. Additionally, the vendor has detailed processing recommendations for every product.

5 TPV – olefin-based TPE, vulcanized, is introduced

This class of polymer compounds containing cross-linked rubber particles makes up a smaller part of the entire world TPE (thermoplastic elastomers) market in (see Chapter 12) but this product's family offers a multifaceted technical property profile. In a closer look, TPV (thermoplastic vulcanizates) are a subsequent development of TPO blends (TPO-M). Generally, polymer compounds change their morphology during a processing step, even more so in a blend where the components are near parity. With more stress in the melt process (without intensive mixing), which means more shear and higher temperature, the different phases tend to separate. This reduces the mechanical performance of the material because the mobile phase is no longer finely dispersed as prior to the processing. Additionally, with an increasing amount of rubber that can become the continuous phase, the thermoplastic is dispersed in it. To avoid that phase inversion, the rubber particles are cross-linked during the compounding step keeping a finely distributed structure as the mobile phase even during subsequent melt processing.

This stable processing behavior is beneficial for extrusion, where a good melt stability is needed. To make bellows, such as those seen in Fig. 5.1, a large soft tube will be enclosed in a mold directly after the extrusion die and then blown with high pressure onto the cold-structured mold wall. This process is called extrusion blow molding and is the primary method for manufacturing bellows.

Fig. 5.1: Bellow from blow molding of TPV (source: Mocom).

According to the ISO 18064, the composition of the material should be classified. Commonly, the expression is a combination of the cross-linked rubber followed by the thermoplastic material:

TPV – (EPDM + PP)   Polypropylene matrix + ethylene–propylene–diene–copolymer elastomer
TPV – (NR + PP)   Polypropylene matrix + natural rubber elastomer

https://doi.org/10.1515/9783110739848-005

The norm offers only a few published variations in the market, in which only poly-propylene (PP) matrices are listed. Meanwhile, other grades are offered, which are based on polyamide (PA) or polybutylene–terephthalate (PBT) matrices providing TPV with higher temperature stability and performance.

5.1 Manufacturing process

The common TPV combination in the market is the blend of PP as the hard phase and ethylene–propylene–diene–copolymer (EPDM) as the flexible phase. The compound is vulcanized during the intensive mixing of the components, predominantly with the presence of ionic species such as phenol–formaldehyde or in a radical process with peroxides or in specific cases with siloxanes. The cross-linking of the rubber phase with sulfur is also possible and common but it is seldom used here due to unpleasant odor. Furthermore, sulfur-based vulcanization requires higher temperature and thereby higher reaction rates that are difficult to regulate. Similarly, electronic beam irradiation provides inferior process control due to a high rate of side reactions.

TPV compounding takes place under high shear in a twin-screw extruder, and the cross-linking agents are added at the proper time and place. This dynamic vulca-nization is also commonplace in a kneader, whereby this is mainly used in a labora-tory or a pilot plant rather than in production scale. The kneader is a discontinuous process in a closed chamber with two rotating kneading blocks for mixing high vis-cous liquids. The shear is given by the geometry and the torque. This process allows more flexibility, but it takes a high effort to produce in big scale and same lot-to-lot homogeneity compared to an extruder (Fig. 5.2).

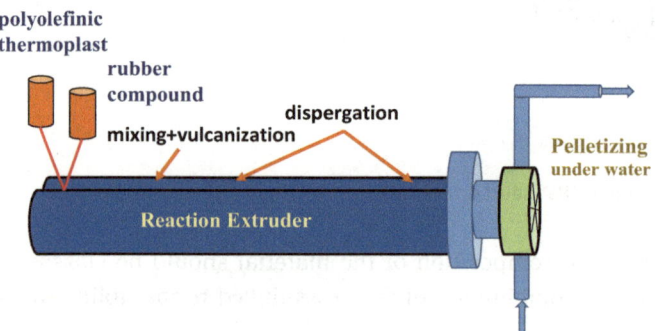

Fig. 5.2: Model of a continuous reaction extrusion process to make TPV.

The rubber component (e.g., EPDM) is fed in the extruder first or simultaneously with the thermoplastic (e.g., PP) intensively mixed. Addition of the cross-linking agents at the right time causes the double bonds of the rubber build-up covalent connections. Noting that only the rubber has unsaturated units, the vulcanization takes place very specifically in the soft phase. Initially, the rubber component is the continuous phase, consistent with a higher volume share. This reverses when cross-linking occurs, and the rubber is dispersed in little pieces, with a phase diameter of down to 1µm. The more evenly distributed the particles are, the better the overall mechanical properties of the material. The optimal size is 1–5µm for the rubber phase in a TPV. The described manufacturing process provides the opportunity to add fillers and oil simultaneously, when not already present in the rubber component.

That process is possible with peroxides to achieve higher cross-linking rate and extended mechanical properties, but it has to be done very carefully. The radical reaction can also occur in the PP matrix which is not desirable. Either with a feeding of the rubber component via side extruder or other specific methods in Fig. 5.3, an example of such a TPV (Mitsui) can be seen in an automotive interior part. That instrument panel has a sublayer of foamed TPV to get a softer touch.

Fig. 5.3: Instrument panel (red) of TPV on a polyolefin body (source: Continental).

Many development activities are ongoing with TPV materials. There are publications about some different thermoplastic matrices like polyamide (PA), polymethylmethacrylate (PMMA), polystyrene (PS), polybutylene therephthalate (PBT), polyetylene therephthalate (PET), and different elastic components including nitrile–butadiene rubber (NBR), natural rubber (NR), butadiene rubber (BR), polyacrylate elastomer (ACM), and silicone rubber (VMQ). Therefore, further designations could be derived from the nomenclature system of ISO 18064:

TPV – (NBR + PA)   Polyamide matrix + nitrile–butadiene rubber elastomer
TPV – (ACM + PA)   Polyamide matrix + acrylate rubber elastomer
TPV – (ACM + PBT)   Polybutylene–terephthalate matrix + acrylate rubber elastomer
TPV – (VQM + TPU)   Thermoplastic polyurethane matrix + silicone rubber elastomer
TPV – (EVM + PBT)   Polybutylene–terephthalate matrix + ethylene–vinyl-acetate rubber elastomer

The new developments should achieve improved temperature stability and a better resistance to environmental influences or certain media like oil and fuel. These properties are mainly influenced by the continuous thermoplastic matrix and have motivated the selection of PAs like PA66 or polyester-like PBT had been selected. Currently, in relation to the information of several suppliers, EPDM + PP is the state-of-the-art material for TPV with a market share of more than 90% and the aforementioned alternatives remain small specialties.

5.2 Properties

As long as a TPV has a nonpolar structure (PP/EPDM), the resistance to low-molecular-weight oils and fuels is weak because the polymer will be soluble. The same holds true for TPV as described earlier for TPO. Against the impact of light irradiation and oxidation, these materials must be stabilized. Water-based media are not critical for them. Therefore, a TPV with a PA or PBT matrix can be more receptive to these, but oil and grease were supposed to be less severe. The matrix is mainly responsible for the temperature stability, even when the cross-linked rubber particles have a stabilizing effect too. It strengthens the material and is reduced accordingly when the continuous plastic phase starts to melt.

In Fig. 5.4, the dynamic mechanical analysis (DMA) is considered again about three different TPVs (for explanation of that method, see Section 11.3).

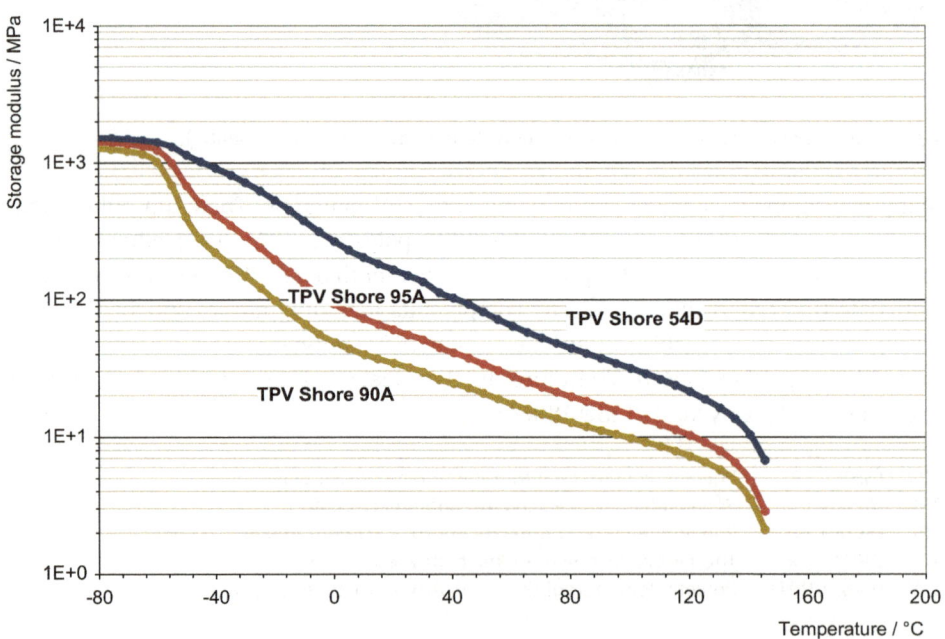

Fig. 5.4: Dynamic mechanical analysis (DMA) of TPV polymers from ExxonMobile (Shore 90A – 3s), (Shore 95A – 3s), and (Shore 98A/54D – 3s); test specimen is in torsion mode at a frequency of 1Hz.

The progression of DMA from a pure PP would have been a horizontal line until the melting point of the polymer. Then, the modulus drops very quickly. Here, with a share of elastomeric particles, the modulus decreases continuously with increasing temperature. The mentioned plateau, characteristic of an ideal cross-linked rubber, is rarely possible with such a blend structure. The higher the temperature is, the more the molecules move, and the boundary between the two phases decreases, which in turn causes a lower modulus. The influence of the elastomeric phase is easily recognized, which means the glass transition is more pronounced with higher content of rubber particles. This character is further enhanced by adding oil, making the material softer and more flexible. This is common practice in the market, where more soft grades of TPV are prevalent, rather than those of a higher Shore A hardness. TPV of Shore D are scarcely available, particularly as such grades would approach the area of much cheaper impact-modified thermoplastic PP. Figure 5.5 shows the DMA of two plasticized TPV grades, where more oil is added compared to the samples above.

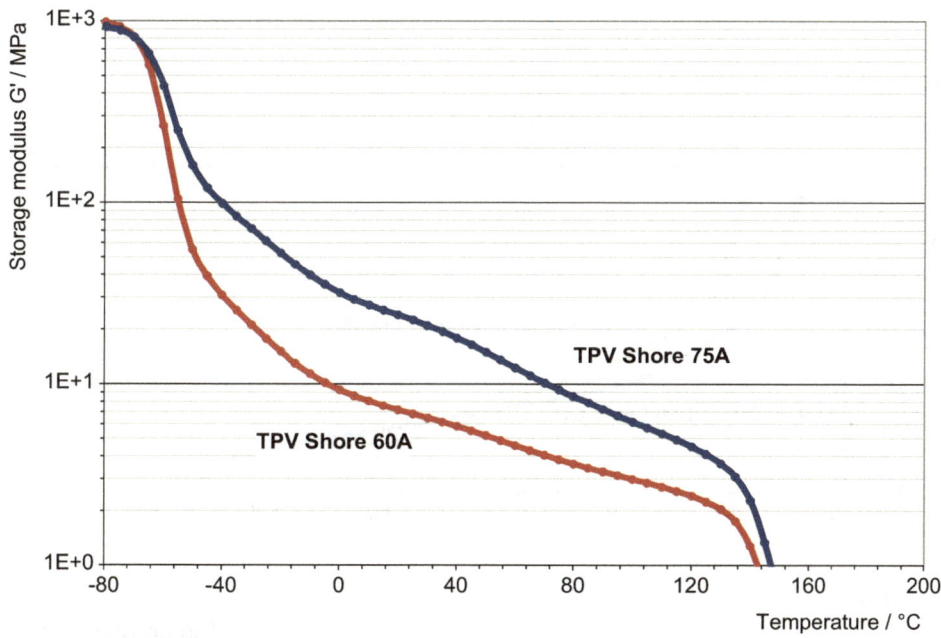

Fig. 5.5: Dynamic mechanical analysis (DMA) of TPV polymers from Mocom (Shore 60A – 3s) and (Shore 75A – 3s); test specimen is in torsion mode at a frequency of 1Hz.

Again, the high content of rubber particles and oil creates a distinct glass transition and decrease in modulus of these soft TPV. Furthermore, at the service temperature, the elastic state is flatter compared to harder grades. A very good cold flexibility can be expected and the modulus progression over the temperature is supposed to be more constant.

Because of their olefinic character, these TPV are a preferred choice as an alternative to rubber materials which provide more applications for softer grades rather than harder ones. The same situation is known for the TPS (see Chapter 6). This relation to the classical rubber materials had been so close that some technical expressions from that trade had been transferred. As already mentioned, the scale of formulations is given in phr (parts per hundred rubber) and not in percentage numbers. To some extent, TPV is still designated as thermoplastic rubber.

The DMA in Fig. 5.6 is a comparison between a TPV of Shore 60A and a styrene–butadiene–rubber mixture (SBR, oil extended, 31.6% carbon black, sulfur vulcanization) of similar hardness. Such a rubber is often used in car tire constructions. The curves generally follow the same progression, which opens the door for a variety of possible TPE alternatives to rubber materials in technical applications. Of course, such a TPE for car tires must be excluded because it will not withstand such heat build-up and dynamic impact.

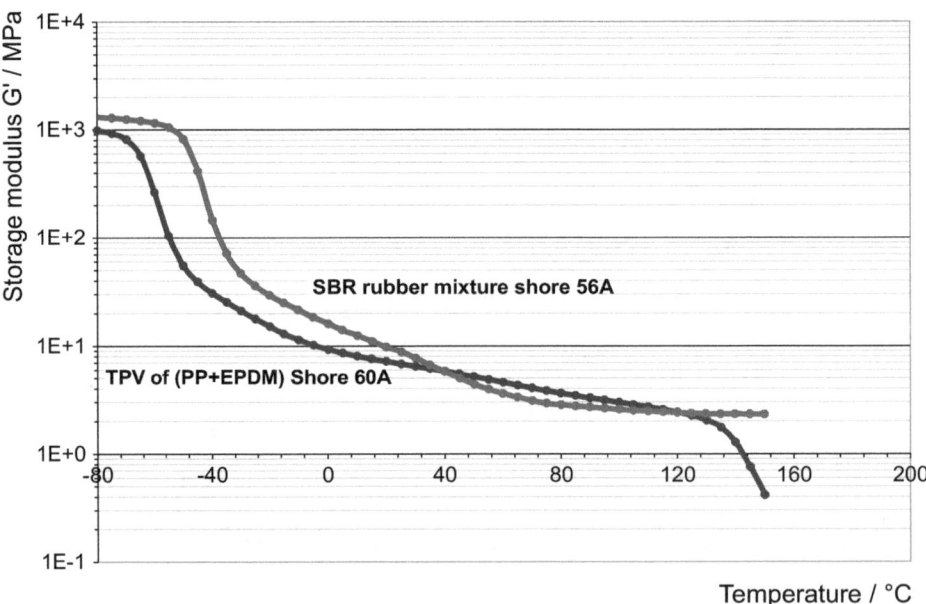

Fig. 5.6: Comparison of DMA of a vulcanized SBR (Shore 56A – 3s) and a TPV from Mocom (Shore 60A – 3s) on a test specimen in torsion mode at a frequency of 1Hz.

All materials, typically used for TPV, are not brittle below 0°C and have an elastomeric character. The modulus breaks down above 130°C, which is governed by the heat stability of PP, the continuous matrix. Until the drop of modulus, the TPV shows a good dimensional stability, an important property to achieve good values, that is, low compression set depending on the formulation and selection of suitable components.

A rough overview of the mechanical properties of the TPV product family by tensile testing (DIN 53504, ISO 37, ASTM D 412) over a hardness range of Shore 30A to 50D:

Tensile strength   2–25MPa
Elongation at break   300–500%

Values of compression set (see Section 11.2) are tested at different temperatures. According to vendor data sheets, compression set for TPV is <20% at room temperature, 20–60% at 70°C, and even at 100°C, such low values are reported in some cases. The main role source of this exceptional heat stability is the cross-linked disperse rubber phase, which is embedded in the thermoplastic matrix. This also provides the good values of compression set, establishing TPV as a common material in the market of gaskets and seals (Fig. 5.7). The wide selection of material combinations offers a wide range of property profiles and driving force behind recent efforts to enhance the elastomeric phase with high-performance engineering plastics.

Fig. 5.7: Extruded gasket for car doors (source: Scholz).

What about the elastomeric properties? In Fig. 5.8, the results of intermittent stress–strain measurements are presented. The progression of the curves shows the degree of deformation with increasing strain on the test bar (for explanation of that method, see Section 11.4).

It can be observed as with other TPE that with increasing hardness, the residual elongation increases, which means the elasticity is reduced. On the other hand, the extensibility of TPV increases with higher hardnesses, opposite to typical rubber behavior. This phenomenon is explained in that with a higher content of elastomeric particles, the overall strength of the compound is reduced because the continuous PP

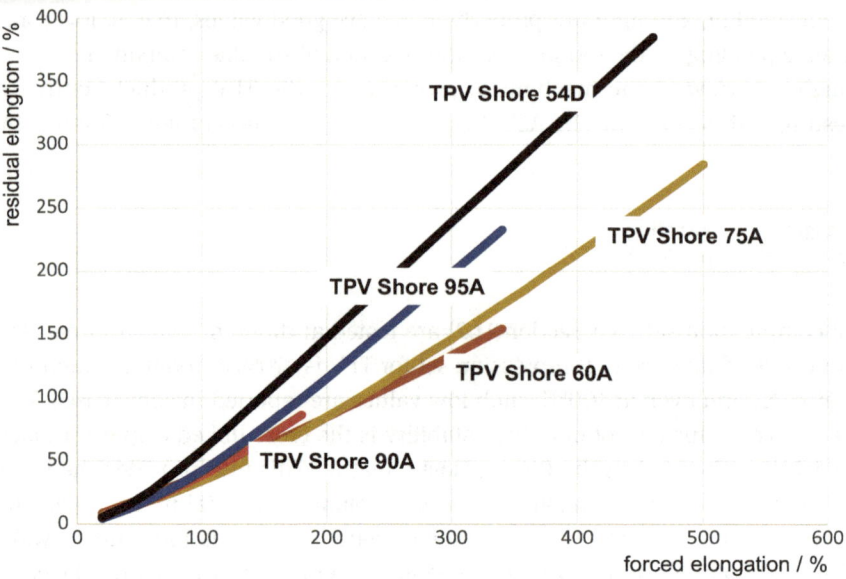

Fig. 5.8: Strain values from the intermittent stress–strain measurement of TPV – (EPDM + PP) from ExxonMobile (Shore 90A – 3s), (Shore 95A – 3s), (Shore 98A/54D – 3s), and from Mocom (Shore 60A – 3s) and (Shore 75A – 3s).

phase becomes thinner and less durable. At a high level of mechanical stress upon extension, the particles from the disperse phase act as point defects, even when they have a small diameter (around 1 μm). This effect can be compensated to some extent by adding oil to help achieve higher elongation. A brief look into the application field of TPV (profiles, seals, and overmolding) shows that a high extensibility is hardly requested.

5.3 Processing

It is possible to make finished parts out of TPV using all the well-known thermoplastic processing methods. Before selecting a new material, the operator should inform himself/herself about its thermal behavior. A well-established characterization of TPE, the melt flow rate (MFR), was also reported and useful for working with TPV. The suppliers, however, have opted out of reporting MFR because the morphology of the material changes during the measurement, negating the value of such a measurement. At low shear under MFR conditions, the structure of TPV still is too inhomogeneous, and it is not reasonable to reliably extrapolate to viscosities at higher shear conditions. It is better to generate melt flow curves of viscosity over shear rate in a rheometer (for explanation of that method, see Section 11.7) and then select the viscosity at a specific shear rate, related to the real conditions at the processor. In Fig. 5.9, an example of a TPV with hardness 75A is portrayed by such a measurement. It is apparent that the processing behavior of that grades differs only slightly despite the different temperatures of 190–210°C.

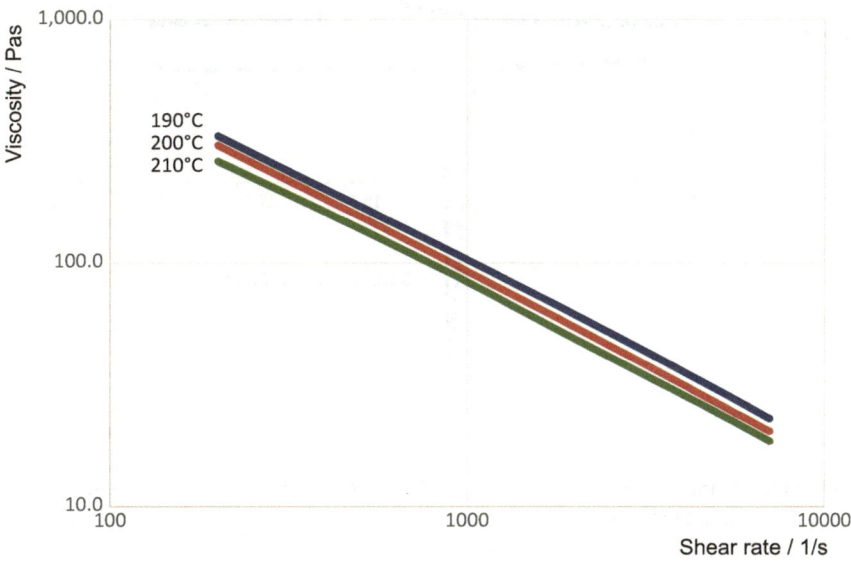

Fig. 5.9: Viscosity curves of TPV – (EPDM + PP) from Mocom (Shore 75A – 3s) in capillary viscosimeter (HKV) at 190, 200, and 210°C.

Herewith a broad processing window is given, which means a low influence of temperature deviations, although it should be taken into account that at high shear rates the PP will accumulate at the surface.

In practice, the manufacturer of TPV uses the spiral flow meter to characterize the melt property of a lot for injection molding application (for explanation of that method, see Section 11.8). The mold is a round or rectangular channel with regular marks to get the information about the flow length easily. In the TPV (similar to TPS) business that procedure is a usual customer service.

Additionally, the pure thermal behavior is of interest for the operator seeking a more targeted temperature profile on the machine. This information can be obtained from differential scanning calorimetry (see Section 11.5).

This figure (Fig. 5.10) shows a relatively hard TPV which displays very clear thermal transitions. The distinct melting peak is driven by the PP phase, which is an indication setoff where to set the processing temperature. The cooling curve shows the temperature at which the material recrystallizes, and the part will solidify in the mold. Both melting and cooling curves are not drastically far away from each other, indicating a short cycle time for the injection molding process.

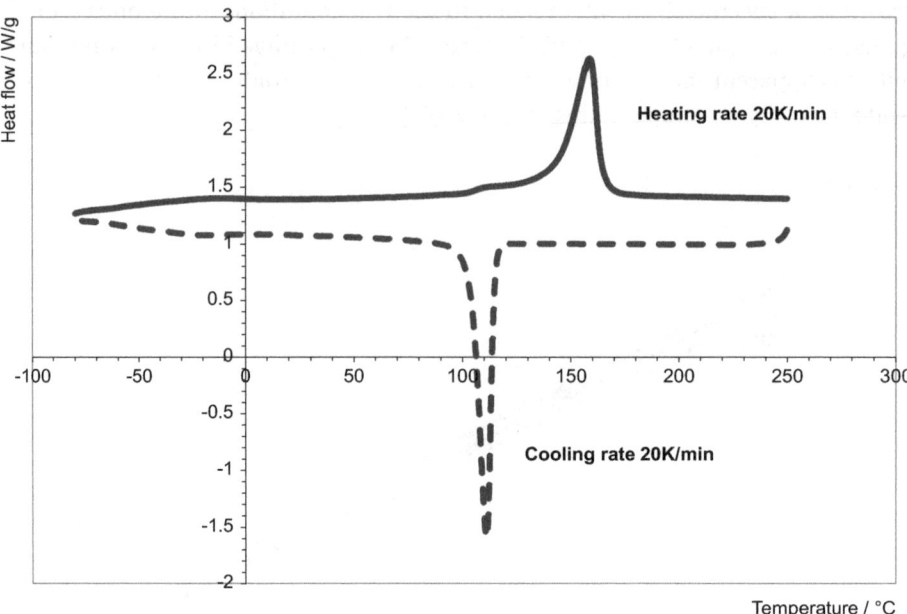

Fig. 5.10: Differential scanning calorimetry of TPV – (EPDM + PP) from ExxonMobil (Shore 98A/ 54D – 3s), at a temperature ramp rate of 20K/min.

TPV composed of PP and cross-linked EPDM remain predominant in the market. They have a nonpolar character causing very low water or moisture uptake. Accordingly, it could be possible not to predry the pellets before the melt processing step. Nevertheless, predrying is recommended by the supplier to avoid any excess moisture caused by higher polarity attributed to cross-linking chemistry. Additionally, the compounders add other polymers to improve adhesion to more polar plastics and such modifications can also increase polarity in TPV. Furthermore, under certain climate conditions, it is possible for water to condense on the surface of the pellets. This causes undesired slippage in the machine and difficulties arise because the material does not homogenize completely. In such a scenario, contacting the material supplier is always recommended, as they provide the best guidance on their products' properties and handling.

6 TPS – styrene-based TPE is introduced

Alongside the biggest group among the TPE (thermoplastic elastomers), the thermoplastic styrene elastomers share the next largest portion of the TPE family (see Chapter 12). On account of its excellent price–performance ratio, this product group is found in use in almost all areas of daily life. This being said, this material can hardly be considered as a pure polymer in technical applications. The styrene polymer has an amorphous structure. It melts very slowly under elevated temperature and this hard phase is very stable. The processability is difficult, so to improve it, an oily plasticizer and/or thermoplastic polymers are added. The typical thermoplastic chosen is PP (polypropylene).

With regard to the pure TPS (styrene block copolymers), it has a triblock construction with an olefinic or aliphatic soft segment between two hard polystyrene blocks. The chemistry of segments is classified in the previously referred to norm ISO 18064:

TPS–SBS   Styrene–butadiene–styrene (not hydrogenated)
TPS–SIS   Styrene–isoprene–styrene (not hydrogenated)
TPS–SEBS   Styrene–ethylene–butylene–styrene (hydrogenated)
TPS–SEPS   Styrene–ethylene–propylene–styrene (hydrogenated)
TPS–SIBS   Styrene–isobutylene–styrene (hydrogenated)

As long as the polystyrene phase is smaller than the flexible one, a spherical structure of the hard segment is preferred firstly because they are amorphous, and secondly, to minimize the interfacial area due to the poor miscibility of the phases. This inverts to the opposite structure when the hard segment has the higher ratio and the flexible segments build up a spherical structure in the continuous phase. This has a definite influence on the character of TPS, at least on the durability and polarity.

The different morphologies of this TPE are very complex and less relevant for the processor or user. With respect to the reality that the more important information about the TPS grade is to know whether the polymer is hydrogenated or not, that is, the nonhydrogenated polymer has double bonds in the chain which are very sensitive against oxidation (see Section 6.1). The new draft ISO 18064 will differentiate between them by TPS-H for the hydrogenated version and TPS-N for the other one in the first place.

6.1 Manufacturing process

In a batch reactor process, the polymer segments are assembled in an inert solvent. The preferred initiator butyl lithium begins a living anionic polymerization, usually with styrene blocks, followed by the polymerization of the olefinic soft segment

https://doi.org/10.1515/9783110739848-006

from butadiene or isoprene, from the reactive ends and ultimately the final styrene segment build-up, creating the third block (Fig. 6.1). Thus, a triblock structure is formed. The ratio of initiator and monomers controls the respective block lengths.

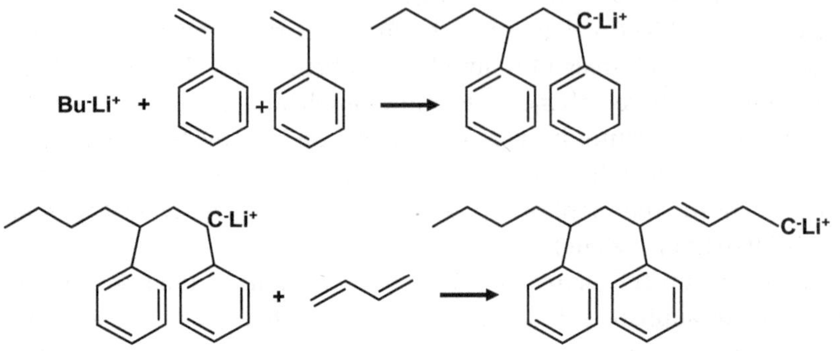

Fig. 6.1: Living anionic polymerization of SBS, initiated with butyl lithium as catalyst on styrene to polystyrene with the living anionic chain initiating polymerization of butadiene in turn.

Butadiene polymerizes from either both or one of its double bonds into a linear chain giving rise to the possibility of a branched vinyl group. This branching increases the flexibility of the soft phase. In most cases, a styrene content of 10–40% is desired to achieve an elastomeric character of the final polymer. The hard phase content has a limited influence on the mechanical properties and mainly affects the processing behavior. Most of these properties are steered by creating the suitable compound, dedicated to the application.

To improve the resistance to environmental influences and improve process stability, the double bonds in the chains should be hydrogenated by more than 99%; in other words, nearly completely. From an SBS an SEBS arises, and from an SIS, an SEPS evolves. The hydrogenation will be done with hydrogen under high pressure and temperature alongside a metal catalyst with a very porous surface. Unsaturated polymers show the tendency to become yellow and to cross-link under the influence of oxygen, which is accelerated at elevated temperature. This is the reason that a compounder uses a hydrogenated version of TPS for technical applications. Especially in an overmolded piece, no changes in color or property are permitted to occur. Overmolding of rigid thermoplastic parts by TPS in a two-component injection molding process is one of the main applications of that TPE family. In such a multicomponent injection molding, mainly, the rigid component will be injected first, the mold opens a bit, and the soft grade will be applied on it via a second screw (Fig. 6.2). As a result of this, a tool gets a smooth grip, a seal can be fitted directly in the manufacturing process, and many more things.

This versatile modification of a TPS enables many opportunities to address market needs with highly specific and adapted technical solutions for the customer. Whether

Fig. 6.2: Screw clamp with rigid body of polypropylene and overmolded with soft TPS (source: Allod).

the compounding or a reaction process of TPE, a twin-screw extruder with parallel rotating double screw is preferred. To achieve the desired properties, this opens a lot of opportunities by versatile screw design, filling rate, shear, and much more. Table 6.1 shows a typical formulation:

Tab. 6.1: Typical formulation of a
TPS–SEBS compound of 60–70 Shore
A hardness (Dynasol).

SEBS	25%
PP	10%
Oil	25%
Fillers	35%
Miscellaneous	5%

Oil has a twofold function: not only does it have a plasticizing effect but it is also an important processing aid. Pure TPS can be melt processed under high energy conditions; in other words, under high shear. The better processing is further aided by addition of PP and oil. Additionally, a filler, often calcium carbonate, helps to improve handling and to reduce shrinkage, which is important when a precise form is requested. Ultimately, the material supplier must adapt the mixture to customer's request including color specification before the product is finalized. The customer purchases this and processes it without the need of further additions.

The simplest and cheapest TPS, made of styrene and butadiene (SBS), is mainly used as is to modify other materials or for short-term applications. The biggest application for SBS is the modification of asphalt to increase damping and to reduce the roll resistance. Another business for short-term use is film for barrier layers in diapers. For pure technical industry, the SBS is hardly suitable related to the said reasons.

6.2 Properties

As previously mentioned, the technical use of TPS is largely successful when compounded together with PP or oil, and TPS performs best in a hydrogenated form. These exist mostly as SEBS or SEPS. The formulation of the mixture has the biggest influence on the resistance to chemical substances. Selection of PP generally boosts this chemical resistance. In general, TPS grades are resistant to polar organic liquids and water-based media (mild acids or bases). This statement holds true for unfilled compounds. For products with fillers, the environmental resistance is also dependent on the kind of filler which can help improve the material durability in an aqueous environment. In unpolar media, TPS compounds tend to swell extensively, which is accompanied by a reduction in mechanical properties. In some solvents like isopropanol, the material starts to dissolve, and a loss of volume occurs. At elevated temperatures, the chemical breakdown is accelerated.

TPS presents its elastomeric properties best in the low hardness range as long as the hard blocks are in the disperse phase (see introduction in this chapter). The hardness of the blends will be adjusted by the ratio of PP and oil, whereas the cold flexibility remains. In Fig. 6.3 of a dynamic mechanical analysis, three compounds with different hardnesses are selected: an SEBS with low PP content and a lot of oil

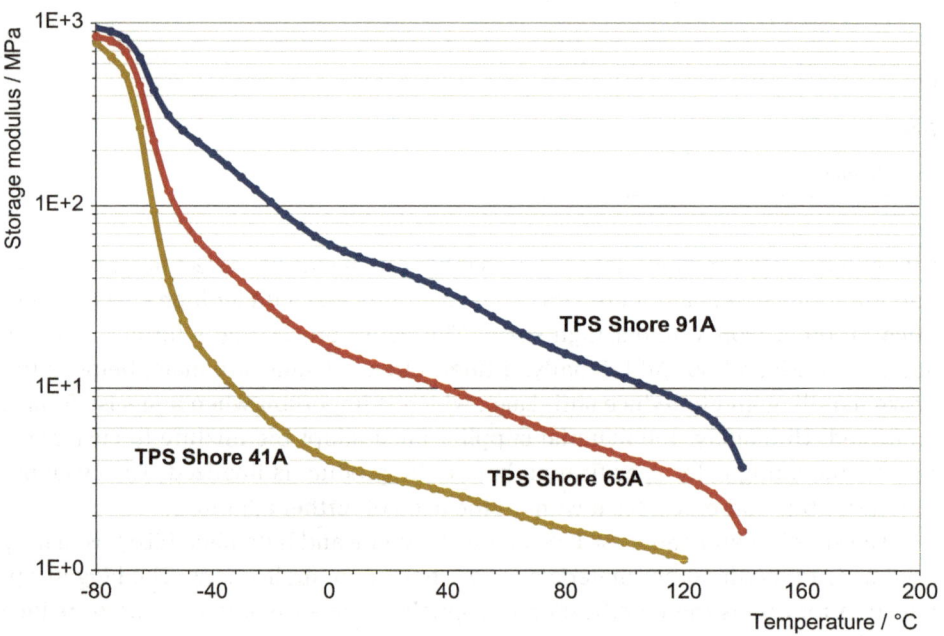

Fig. 6.3: Dynamic mechanical analysis of TPS compound of Allod (Shore 41A – 3s), (Shore 65A – 3s), and (Shore 91A – 3s) on a test specimen in torsion mode at a frequency of 1Hz.

(Shore 41A), a medium PP and oil content (Shore 65A), and one with higher PP content and a less oil (Shore 91A) (for explanation of that method, see Section 11.3).

It is easily visible that the content of PP increases the modulus as well as the heat resistivity, which is close to the service temperature of the grade. This means that the harder the material, the higher the breakdown temperature. Over the whole progression, the curves fall continuously; in contrast to a soft rubber, neither a deep glass transition nor a subsequent flat plateau of modulus is seen. This is influenced by the increasing amount of PP, even when the amorphous hard phase of styrene blocks is well distributed. With increasing temperature, subtle changes in morphology of the blend should be considered. Nevertheless, there are still good mechanical properties at service temperatures, also adjustable by the amount of PP. Cold flexibility does not suffer as a consequence of the PP addition. Moreover, the oil has a plasticizing effect due to its compatibility with the soft phase. It increases the mobility of the chains in the soft phase and the glass transition thus moves to lower temperatures. Looking at the curve of the softest sample, the deep glass transition can be seen very well. In general, these three examples illustrate the versatility of TPS mixtures and the opportunities of a compounder to respond to specific customer requests.

As shown in thermoplastic vulcanizate (TPV) chapter, it is worthwhile to compare the DMA of a TPS with a certain rubber mixture. Figure 6.4 illustrates a soft TPS-grade

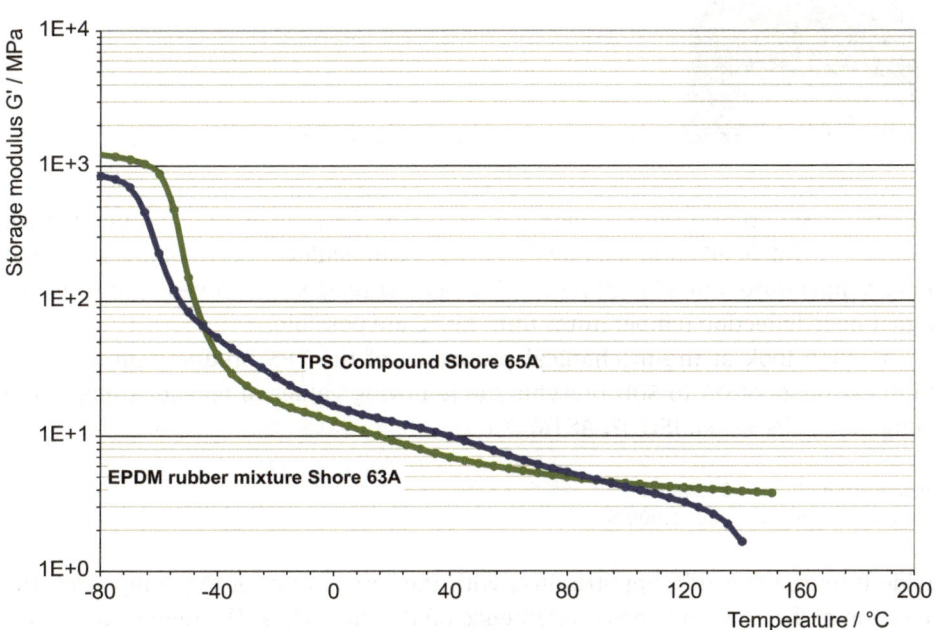

Fig. 6.4: Comparison of dynamic mechanical analysis of a TPS from Allod (Shore 65A – 3s) and a rubber formulation of ethylene–propylene–diene–rubber (Shore 63A – 3s) on a test specimen in torsion mode at a frequency of 1Hz.

Shore 65A with a rubber mixture based on an ethylene–propylene–diene–rubber (EPDM, 38% carbon black, 24.2% oil, sulfur vulcanization) and a hardness of Shore 63A. Such an EPDM rubber is a typical formulation for damping elements.

Both materials are used for gaskets and seals, and as depicted by DMA, there is a very good overlap of TPS with the EPDM rubber over a wide range of temperature. At high temperatures, the TPS transitions to the melting phase and the EPDM is still stable because of its chemical cross-links. Figure 6.5 shows a typical application as a seal in the area of food and beverages. Ease of manufacturing and ability to color justify the value of producing such a part from TPE.

Fig. 6.5: Seal ring of TPS for swing tops (source: Allod).

One important criterion to evaluate the suitability for seals is the compression set (ISO 815, ASTM D 395) and usually TPS grades can achieve values of about 20% at room temperature and 30% at 70°C. With special recipes, improved values means higher heat deflection temperatures until 150°C are possible.

A rough look at the mechanical properties of the TPS product family over a hardness range of 0A to 50D presents the following values in tensile strength and elongation (DIN 53504, ISO 37, ASTM D 412):

Tensile strength   2–25MPa
Elongation at break   200–1,000%

Aside from the versatile opportunities with the formulation of the compound, the selection of the TPS grade has an influence on the properties. The higher the molecular weight of the TPS together with a suitable amount of PP and filler, the better the heat stability and the creep resistance. This is very helpful for high-performance sealing materials. When a TPS is not completely hydrogenated, the residual double bonds in the chain can be used for cross-linking after the molding process to get a higher heat stability. There appear to be only few grades in the market for this use.

Now let us go to the elastomeric properties. Again, they are illustrated by the intermittent stress–strain measurement (for explanation of that method, see Section 11.4).

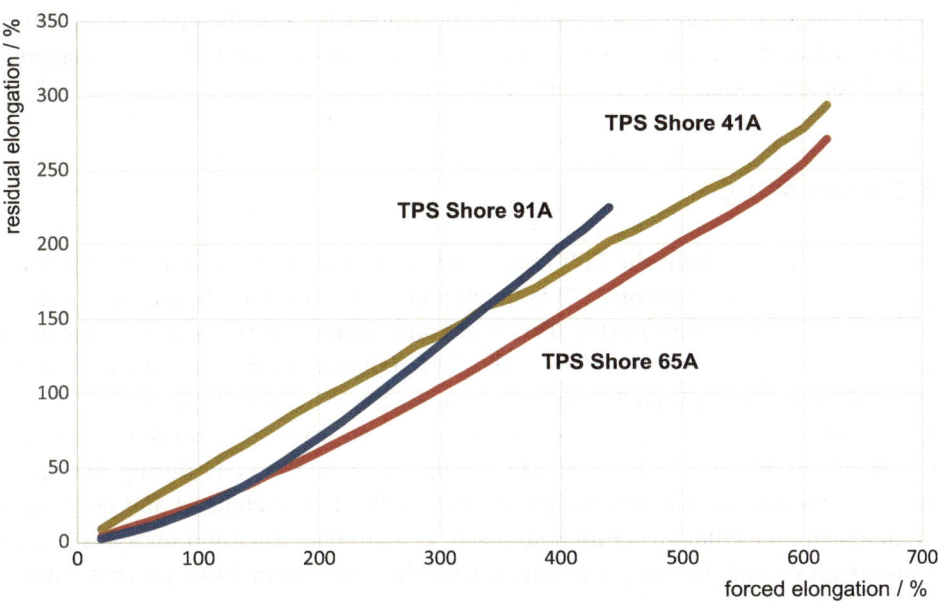

Fig. 6.6: Strain values from the intermittent stress–strain measurement of TPS compounds from Allod (Shore 41A – 3s), (Shore 65A – 3s), and (Shore 91A – 3s).

Because of the different influences of the PP and the oil combination, the results are normally diverse, whereas the selection of materials in the diagram (Fig. 6.6) shows that the elastic behavior under increasing elongation is isolated from that. The plots of the soft grades have a nearly linear progression, which is mainly driven by the high content of plasticizer. The nonlinear character of the harder type is due to the high PP content. The decreasing interaction of PP and TPS phase makes the material weaker under high stretch. Here as well, the good flexibility at low elongation is a high value for TPS compounds. Since it is expected that the main application is a combination with rigid thermoplastics by overmolding, the softness, nice grip, and directly molded sealing are the desired properties, rather than a high elongation, as we have seen in Chapter 5 already.

The nonhydrogenated grades (SBS and SIS) are visible in the market only in specific applications, even if they are big ones. The most single use of SBS, the copolymer of styrene and butadiene, is the modification of asphalt, as already mentioned. SBS reduces noise, increases the lifetime, and creates less abrasion of the car tire. Another application is for low price shoe soles where the poor light stability has a lower relevance. Short-term use is an elastic film for diapers, in which the material is elastic enough and a low-cost solution. The SIS copolymer made up of styrene and isoprene is often applied in the adhesive industry and hardly seen as a technical polymer, because they have a low viscosity and therefore poor mechanical properties. As mentioned, the nonhydrogenated version is not used in technical areas because the unsaturated bonds in the hydrocarbon chain are responsible for a weak weathering resistance compared to the hydrogenated ones (SEBS and SEPS).

6.3 Processing

In the last chapter about the TPV family, it was explained that results from a complete flow curve (see Section 11.7) is much more helpful rather than a simply measured melt flow rate (MFR) value. These diagrams show that TPS or TPS compounds have very good melt stability during the thermoplastic processing, comparable to TPV. Additionally, TPS need a high shear to ensure a homogeneous melt. For that reason, an evaluation of processing behavior based on MFR is senseless because the shear rate is way too low and TPS change their morphology during that measurement procedure. It is not completely molten in the test channel and deviations in the macromolecular structure take place. TPS are manufactured for specific customer requests mainly and predominantly for the injection molding process. Subsequently, the characterization of the processing behavior will be done in a spiral flow injection (for explanation of that method, see Section 11.8). This is close to the practice and serves as a reliable information for the processor.

The main technical market for TPS is injection molding, especially multicomponent molding. In that case, a more practical method is used to evaluate the flowing properties of the TPS melt. This is the spiral flowmeter in injection molding, where the flowing behavior can be visualized very well and close to the application. In addition to the flowing behavior, the thermal properties are an important factor. In Fig. 6.7, three different TPS compounds are illustrated in a thermal analysis (DSC; for explanation of that method, see Section 11.5). The processor receives some information about the melting and solidification behavior and can use this information for the settings on the injection molding machine.

The styrene units have mainly an amorphous character and they soften over a broad temperature range. This causes a high viscosity melt with some difficulties in processing the material. This is the reason for making a compound for technical applications with PP and a plasticizer. It is seen from the curves of the samples at high temperatures that PP is part of the compound because the peaks are close to each other, and the area below the peak gives the hint about the content of PP. This area

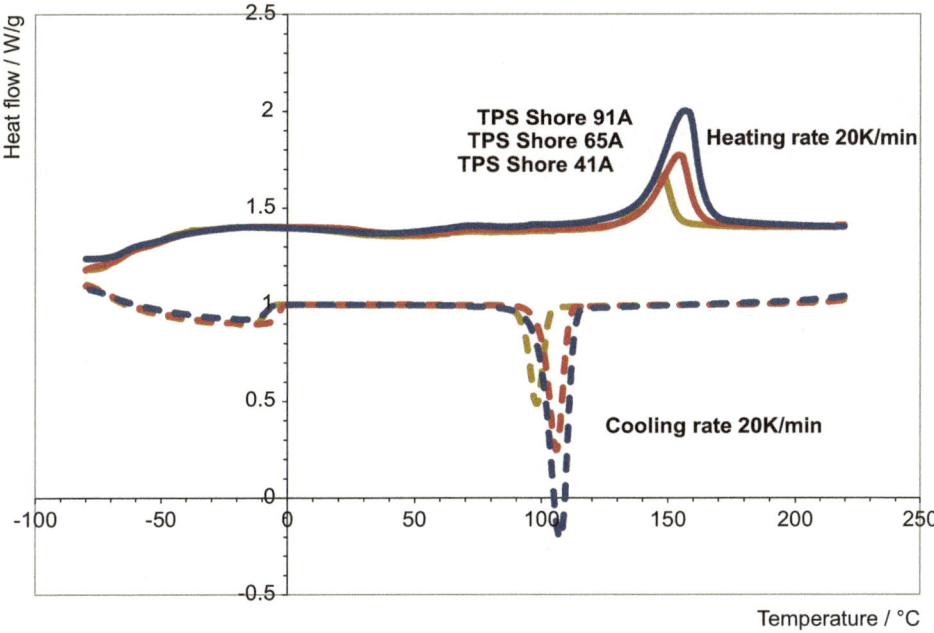

Fig. 6.7: Differential scanning calorimetry of TPS compounds from Allod (Shore 41A – 3s), (Shore 65A – 3s), and (Shore 91A – 3s), at a temperature ramp rate of 20K/min.

represents the heat content (enthalpy) of the component and the higher the content, the bigger the peak is. The softening of TPS itself can hardly be seen. The recrystallization temperature areas are similar for all of the samples. The softest one with a high oil content solidifies at a slightly lower temperature. The peaks of the cooling curve are very narrow, which exemplifies a fast freezing and a good processability.

7 TPU – urethane-based TPE is introduced

Thermoplastic polyurethane (TPU) is a medium-scale player in the group of thermo-plastic elastomers (TPE; see Chapter 12). People chose this material when a high wear resistance is needed while maintaining a good elasticity. The way this is possible will be presented in this chapter.

There are hardly any technical market fields, where TPU is not present. One of the applications that well illustrates the versatile profile of this material is the cable jacketing for automotive applications, for example, the electronic steering system for ABS (anti-lock brake system) and ESP (electronic stability program). These cables (Fig. 7.1) are exposed to many different environmental influences like temperatures from −40 to 80°C in dry and wet atmospheres. Anything present on the street can attack the material, be it mud or salt, even oil and fuel. Furthermore, mechanical impacts of any kind can happen during a car ride.

Fig. 7.1: Cable jacketing for electronic steering system of TPU (source: BASF).

The urethane will be built by a simple fast reaction of an isocyanate (NCO) and an alcohol (OH) without creating a by-product. At the end of every molecular unit is a respective reactive group and the polyaddition reaction starts. Three components come together to react: a short-chain diol with an isocyanate to build up the mainly crystalline hard phase and a long-chain diol again with an isocyanate to create the flexible soft phase (Fig. 7.2).

$$R^1\!-\!N\!=\!C\!=\!O \quad + \quad HO\!-\!R^2 \quad \longrightarrow \quad R^1\!-\!N\!-\!C\!\!\begin{array}{c}\nearrow^O\\\searrow_{O-R^2}\end{array}$$

Fig. 7.2: Reaction to urethane.

https://doi.org/10.1515/9783110739848-007

The chemistry of the segments is classified in the nomenclature of the already mentioned standard ISO 18064:

TPU–ARES   Aromatic polyester–urethane
TPU–ARET   Aromatic polyether–urethane
TPU–AREE   Aromatic polyester–ether–urethane
TPU–ARCE   Aromatic polycarbonate–urethane
TPU–ALES   Aliphatic polyester–urethane
TPU–ALET   Aliphatic polyether–urethane

7.1 Manufacturing process

In general, the structure of a TPU can be described in Fig. 7.3 of a TPU reaction:

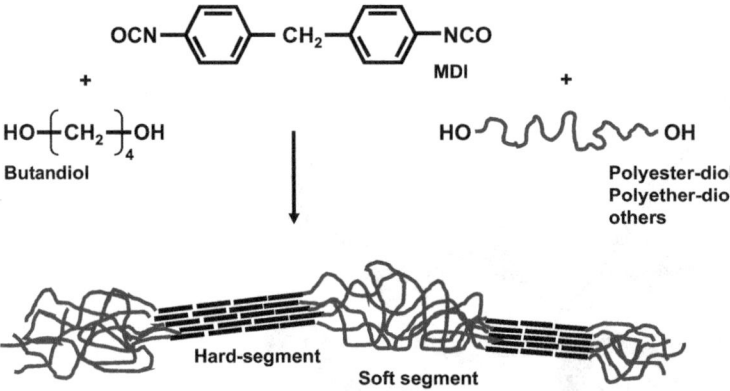

Fig. 7.3: Reaction model of TPU.

Here, the reaction shows an example of methylene diphenyl diisocyanates (MDI) with butanediol as the short-chain alcohol and different polymer diols as the long-chain alcohol. The reactants are mixed intensively in a liquid state, usually at elevated temperatures. The reaction starts spontaneously, and the temperature increases continuously. The isocyanate builds the crystallizing hard segment together with the short-chain diol, with the long-chain one as the flexible soft segment. The polyaddition runs asymptotically to 100% consumption but it ideally needs a posttreatment depending on the formulation, normally of few hours or several days for very soft grades. The reaction product, or finished polymer, should become solid under a minute so that it can be pelletized, dried, and stored. Different methods are possible.

In a simple discontinuous cast process, small batches can be made by heating up the raw materials to keep them liquid, mixing them intensively and once the reaction starts the mixture can be poured onto a heated table. The whole system remains on this table until the temperature reaches a maximum. Curing occurs in a conditioning oven until the material is solid enough to cut, grind, or mill it into a feedable chip or crumb. Optionally, the crumb can be further transformed in an extruder to obtain more regular and homogenous pellets.

In a large-scale continuous manufacturing process, two methods are established. On the one hand, the reaction takes place in a heated tunnel, the so-called belt line (Fig. 7.4). The reactants will be premixed in a mixing vessel and directly poured onto a heated conveyor belt. At the end, the polymer slab must be cooled to enable granulation of the fresh material.

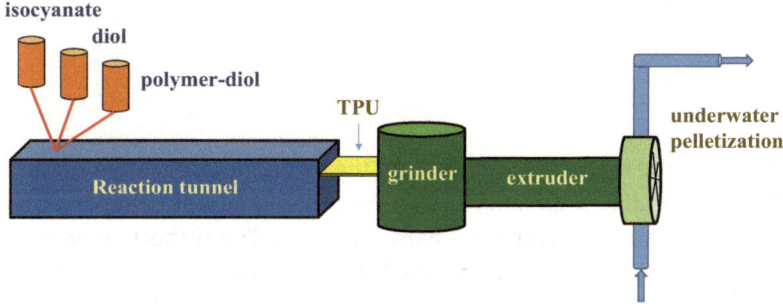

Fig. 7.4: Belt line process for manufacturing TPU.

From the slab, cubic "dice" can be cut continuously, whereas a subsequent transforming step on an extruder is possible to receive a smoother and more free-flowing pellet shape. This occurs in a separate step or directly afterward in a connected extruder as illustrated in Fig. 7.4. Here, the polymer slab exits the line after a cooling zone and goes into a shredder where the chopped pieces are melted in a connected extruder and pelletized. For TPU, an underwater pelletization is recommended, especially for soft grades.

A newer modern process enables the polyaddition on a twin-screw extruder where the liquid reactants can be mixed very well and during the whole reaction step. Since the synthesis takes place in an extruder, it is called reaction extrusion (Fig. 7.5). We learned that expression from the thermoplastic vulcanizate technology (Section 5.1) but the difference is a vulcanization process there rather than a polymer synthesis.

At the end of the reaction extruder, the melt has a viscosity which is high enough to cut the material in an underwater pelletization (usually) getting smooth oblong pellets. Both processes have unique advantages, especially when TPU with different crystallinity should be produced. One could imagine that in a continuously mixing reactor, more transparent polymers are synthesized and on a belt line with no stirring during most of the reaction time, more crystalline ones are produced. The raw materials are not very miscible and crystalline superstructures can be built up without any influence from stirring.

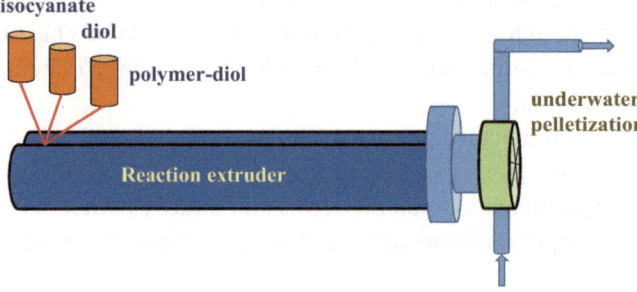

Fig. 7.5: Extrusion process of manufacturing TPU.

7.2 Properties

TPU typically offered in the market are divided, on the one hand, into polyester and polyether based on the flexible segments, and on the other hand, into aromatic and aliphatic TPU related to the semicrystalline hard segments. The respective nomenclature in ISO 18064 has already been mentioned. The following more comprehensive description of the soft segments leads to the fact that for the thermoplastic copolyester (Chapter 8) and the thermoplastic polyamide (Chapter 9), the same circumstances of these flexible elements count. For these polyester and polyamide elastomers, the similar polymer diols are used.

To start with the soft phase, the significant difference is seen in the resistivities of the resulting TPU. A polyester is synthesized from an alcohol and an acid to form an ester and water. On the other way around, when the ester reacts with water back to an alcohol and acid, there will be a cleavage of the polymer chain at the polyester molecule (hydrolysis), subsequently with a drop of mechanical properties. Additionally, that reaction is catalyzed by acid, which means it is an autocatalytic process. Therefore, the formulation often contains a stabilizer to catch the acid which helps to prevent or retard the hydrolysis reaction. This is the reason that polyester-based TPU can be used for many years at ambient temperatures without any degradation.

For an application in a humid and warm environment, it is recommended to use a polyether-based TPU which is inherently stable against hydrolysis and microbe attack. It must be oxidatively stabilized because oxygen radicals attack the polyether chain, and peroxides will be created which are not stable. The cleavage of polymer chain causes bad mechanical properties always. With the proper antioxidants, this TPU can be used many years indoor and outdoor.

Figures 7.6 and 7.7 illustrate the results of long-term studies on the hydrolysis and heat aging of polyester- and polyether-based TPU at different temperatures. It is an Arrhenius plot, which has a criterion requiring the final TPU sample to exhibit 20MPa in tensile strength. That means, TPU samples had been stored under selected atmospheres for a long period of time. In relation to a certain schedule, a sample was taken out and the tensile strength detected. When 20MPa was achieved, the related time was set as the end of good mechanical properties. The logarithmic plot of the time of storage over the reciprocal temperature leads to linear curves which can be used for extrapolation to the lifetime of the tested material.

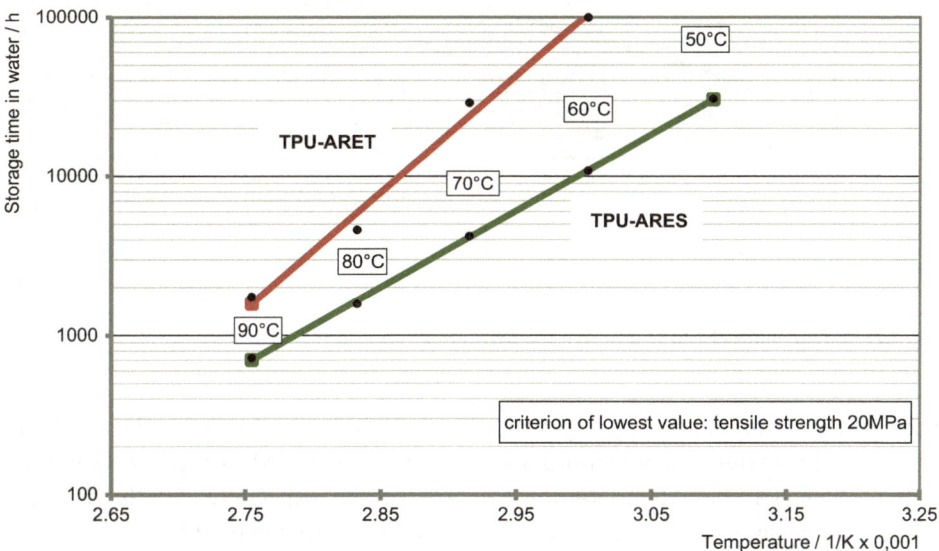

Fig. 7.6: Storage in water of TPU at different temperatures.

It can be seen that the curve of the TPU with polyether soft segment lays above the line of the polyester grade significantly. This relates to the previous explained better resistance against hydrolysis of a polyether. Following the line to lower temperatures, the long lifetime of TPU–ARET in water is obvious. With increasing temperatures, the curves of polyester- and polyether-based polymer begin to converge and meet at 90°C. Considering that oxidation occurs more readily at elevated temperatures, even in water, it is reasonably expected that an ether chain will be susceptible to an oxidative attack. Looking at the results from storage in air (Fig. 7.7), the curves approach with decreasing temperature until they cross, even with a better stability of a polyester TPU against oxidation. But the lower the temperature, the higher the humidity in the chamber is and hydrolysis of the polyester occurs.

In respect to a long period of time in hot air, the curve of the polyester TPU lays above the line of the polyether one. The lines meet at 100°C and the hydrolysis of the polyester segment begins. In case of a request, where a good stability against high temperatures is given, a TPU–ARES is the right choice. It has to be taken into

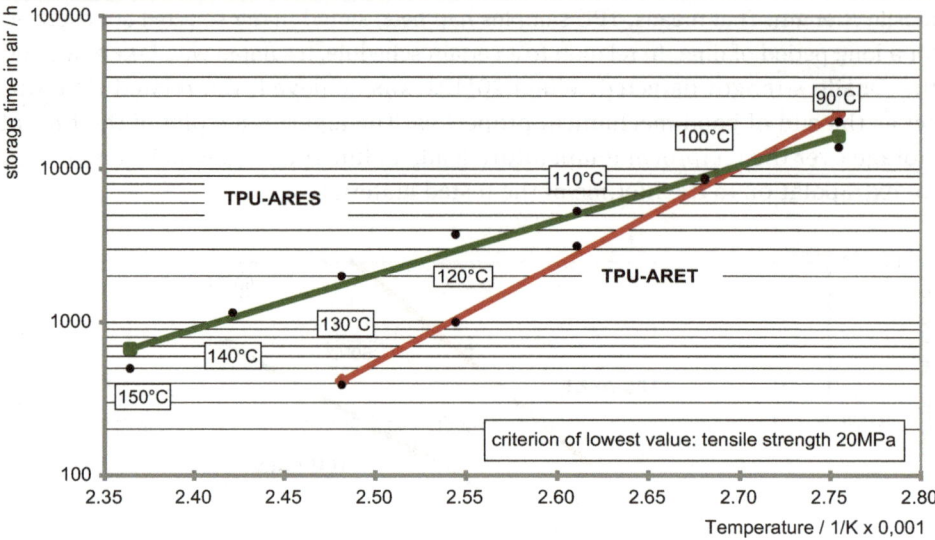

Fig. 7.7: Storage in air of TPU at different temperatures.

account that in an outdoor application, always humidity and heat can be expected, and the right material is a compromise, which is illustrated in Fig. 7.7. Overall, a long lifetime at ambient temperatures can be expected, especially with the right stabilization package. An Arrhenius plot is a suitable calculation on that.

Let us go with hard segments of TPU. In regard to the semicrystalline hard phase out of a diisocyanate and a short-chain diol (see Section 7.1), we define two different groups, the aromatic ones that contain benzene segments and the aliphatic TPU which do not contain aromatic moieties. This relates to the nomenclature in ISO 18064 by AR for the aromatic and AL for the aliphatic grades. Typical representatives are the aromatic MDI (see Fig. 7.3) and the aliphatic hexamethylene diisocyanate (HDI), which is completely linear structured.

What are the relevant differences between aromatic and aliphatic TPU? The crystalline segments based on MDI and butanediol have a very stable structure which gives TPU a high mechanical strength. The flat aromatic rings are adjusted to each other very well, supported by hydrogen bonds which also stabilize the structure. Aromatic rings are normally sensitive to UV light. Upon irradiation, an MDI-based hard phase starts to become yellow without losing any physical performance. In other words, aromatic TPUs discolor under UV light over the time but they keep their mechanical strength. On the contrary, aliphatic TPUs remain colorfast UV light but they require stabilization to prevent degradation. This high resistance to discoloration is capitalized on in automotive interiors to maintain an aesthetic over a long period of time. The example in Fig. 7.8 presents the surface of an interior door handle. A thin

Fig. 7.8: Door handle in automotive interior of soft TPU on a rigid polycarbonate/
acrylonitrile–butadiene–styrene body (source: BASF).

TPU layer was applied in a two-component injection molding process on a glass-fiber-reinforced PC/ABS (polycarbonate/acrylnitril-butadiene-styrene copolymer) body.

In addition to the color stability and the ease of manufacturing compared to a laborious painted applique, this piece has a pleasant haptic. The perception of touch is a subjective factor but it is a very important element in such an application. Furthermore, this TPU is resistant against skin lotions and sweat. In long-term application outdoors, aliphatic TPU should be the preferred alternative, especially when continuous light irradiation takes place and discoloration is not acceptable.

There is an additional aspect, which supports the high wear resistance of TPU. It was explained that isocyanates react rapidly with alcohol groups. Likewise, they build chemical bonds with existing urethane groups, called allophanates. This is a kind of cross-linking, which provides strength but cleaves under processing conditions at high temperatures in the melt and rebuilds afterward due to their reversibility. The high tensile strength at large elongations are of particular advantage in engineering applications. Figure 7.9 illustrates equipment with pneumatic tubes that should maintain a high internal pressure and should be flexible enough still that no kinking occurs when the tube is sharply bent. Dimensional stability during the manufacturing of such a tube is an obvious prerequisite.

All the TPE can be found as a material for tubes in every kind as long as flexibility is desired. Therefore, for all requests in that market TPE are available for engineering, conveying, medical use, and in food technique.

Essential mechanical properties are the values of tensile strength (DIN 53504, ISO 37, ASTM D 412) and the abrasion loss testing (ISO 4649, ASTM D 5963) which are relevant for the interaction of elasticity and strength. Broadly over the product portfolio, the hardness range of nonreinforced TPU is about Shore 50A to 85D with a property range:

Tensile strength	10–60MPa
Elongation at break	100–1,000%
Abrasion loss	20–100mm^3

Fig. 7.9: Pressure tubes of TPU (source: BASF).

Aside from testing the mechanical properties, the semicrystalline hard segments cannot be easily characterized. Often the thermal analysis (DSC, differential scanning calorimetry) is preferred because the thermal transition of a material can be identified very accurately. In the case of TPU, caution is recommended because the morphology of TPU changes during the measurement. On the resulting spectrum, a modified material is represented. For this reason, this method does not hold as much value for TPU compared to other TPE. To avoid that effect, often a second run of a sample will be taken to get a better comparison between TPU, but then the material has a different thermal history than the virgin one.

If it is desired to extend the property portfolio, it is necessary to use suitable additives. Very soft formulations with shore hardness lower than 60A normally contain plasticizer. Without any plasticizer, a low crystalline content of TPU is very sticky and handling the material requires great effort. On the other extreme, obtaining a TPU with a high modulus of about 10,000MPa can only be achieved by adding glass, carbon, or mineral fibers in an extrusion process.

Even here, the dynamic mechanical analysis (DMA; for explanation of that method, see Section 11.3) will be consulted to describe the behavior over a temperature range. Figure 7.10 shows how a TPU formulation can range from a soft elastic material, akin to a rubber, to a rigid thermoplastic in character, although it remains elastic. In an impact test at freezing temperature, it can be seen that such a TPU does not fracture at −30°C. Despite this, the DMA barely presents a distinct glass transition for the hard TPU grades.

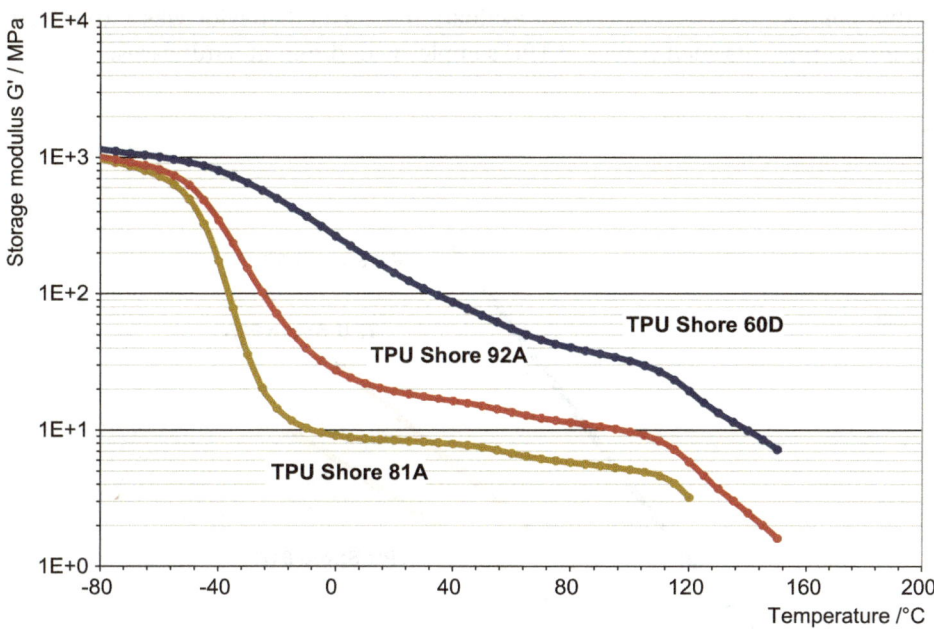

Fig. 7.10: Dynamic mechanical analysis of TPU–ARET of BASF (Shore 81A – 3s), (Shore 92A – 3s), and (Shore 60D – 3s) on a test specimen in torsion mode at a frequency of 1Hz.

As depicted in DMA, the left part of the curves gives an answer on the behavior in the cold. Especially the soft grades show a deep glass transition and a distinct plateau will be shown for soft rubber grades. This character combined with a good wear resistance gives a reason why TPU has a broad distribution in shoe application, in sport: casual or safety shoes. The decrease in storage modulus at the end of the curve signifies the limit of the service temperature. Additional helpful information about the operating temperature during use can be gained from compression set testing (see Section 11.2). It was already mentioned that TPE exhibit creep significantly earlier and easier compared to a cross-linked rubber, and from these values the deflection temperature of two different materials can be compared for evaluation of a possible substitution. Aromatic TPU of moderate hardness have compression set values of about 20% at room temperature, and around 50% at 70°C. Specialty formulations can achieve 30 % at 70°C or 50% at 100°C, an area in which most standard TPU can no longer be measured.

TPUs normally have a polar character and therefore they are not compatible with oils and grease. This causes a good stability against these substances. Basic or acidic media, such as those originating from renewable resources, are generally a weakness for TPU.

High elasticity with good strength of TPU has already been mentioned. At the beginning of the book, the intermittent stress–strain measurement (for explanation of that method, see Section 11.4) to evaluate the elastic behavior was introduced. Figure 7.11 presents such curves of TPU samples with different hardnesses.

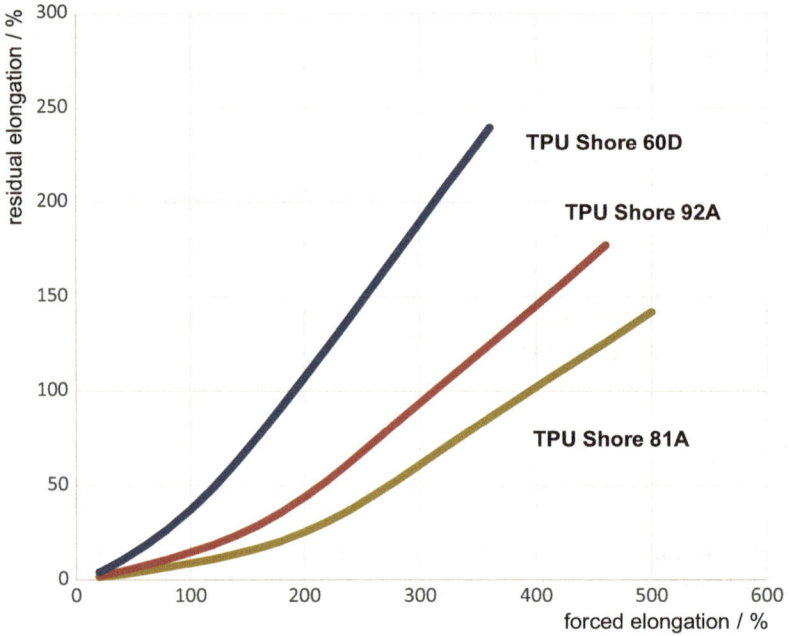

Fig. 7.11: Strain values from the intermittent stress–strain measurement of TPU–ARET from BASF (Shore 81A – 3s), (Shore 92A – 3s), and (Shore 60D – 3s).

The softer the TPU is, the later the deformation accelerates. In particular, the softest grade has a flat curve of residual elongation at the beginning at 100% and higher. This indicates a high elasticity without the Mullins effect, which was introduced previously. With increasing content of hard segments, the stretchability is reduced and deformation upon strain is enhanced. It is not so relevant in practice because the durability of the softer grades is good enough and resembles a rubber material more closely than harder TPU grades. In Chapter 3, which reviews the principle comparison of TPE to rubber, TPU was the representative example and thus will be shown there in more detail.

7.3 Processing

TPU is the most sensitive material in the TPE family with respect to the processing of pellets. In such processing, the polymer melts in the machine and the urethane bonds start to cleave due to the reverse reaction at the ceiling temperatures. The addition of isocyanate and alcohol is an equilibration reaction which derives at cleaving temperature back to the raw materials. It does not revert completely to monomers, but the melt flow character is influenced. This happens consistently in the standard processing conditions of extrusion and more intensively during injection molding because the shear rate is higher. Directly after the processing step, the polyaddition starts again and the molecular weight of the TPU recovers. Therefore, one must ensure that the shear in the melt is not too strong, which means the pitch of the screw should not be too deep, the flow areas are open, and no dead spots are present in the machine. On a twin-screw extruder, any reverse screw element should be avoided because their impact is too severe. Therefore, a constant temperature profile during the process must be kept. All these factors are the reason for a small processing window compared to other polymer materials. The processing of TPU is a longer learning procedure and in every case the vendor recommended guidelines should be followed.

In Fig. 7.12, viscosity curves taken from a high-pressure capillary viscometer (HKV, for explanation of that method, see Section 11.7) are illustrated. With the aid of the core pressure in the melt runner, the pressure is adjusted, and the selection of the die geometry dictates the velocity. Subsequently, the shear stress is determined from the pressure and the shear rate is determined from the output of the melt. The temperature is fixed for every test cycle. The resulting melt viscosity as a function of the shear rate at different temperatures elucidates the significant difference in flow behavior even in temperature steps of only 10°C. This means that during TPU processing, an accurate and constant temperature must be maintained to enable a homogeneous flow behavior, as the processing window is relatively small compared to other thermoplastic materials.

At this point it is worth to emphasize that the plastic database "Campus" contains a huge number of data and diagrams to help a customer for the selection of a polymeric material. Literally, engineering plastics are listed but more TPE will be incorporated.

The aforementioned degradation during processing makes critical that TPU is dried correctly before the melt processing. Vendors strictly recommend a water content of 0.01–0.02% as measured by a moisture vapor analyzer. When the reverse reaction creates the reactive chain ends of isocyanates in the melt, the rebuilding of polymer chain in the presence of water is interrupted because water can react with isocyanate like alcohol groups present in the melt. Thus, when processing moist material, the expected molecular weight cannot be achieved, and ultimate mechanical properties are poor.

Nevertheless, experienced operators are prepared for these circumstances when working with TPU; otherwise, they would not achieve the desired performance of

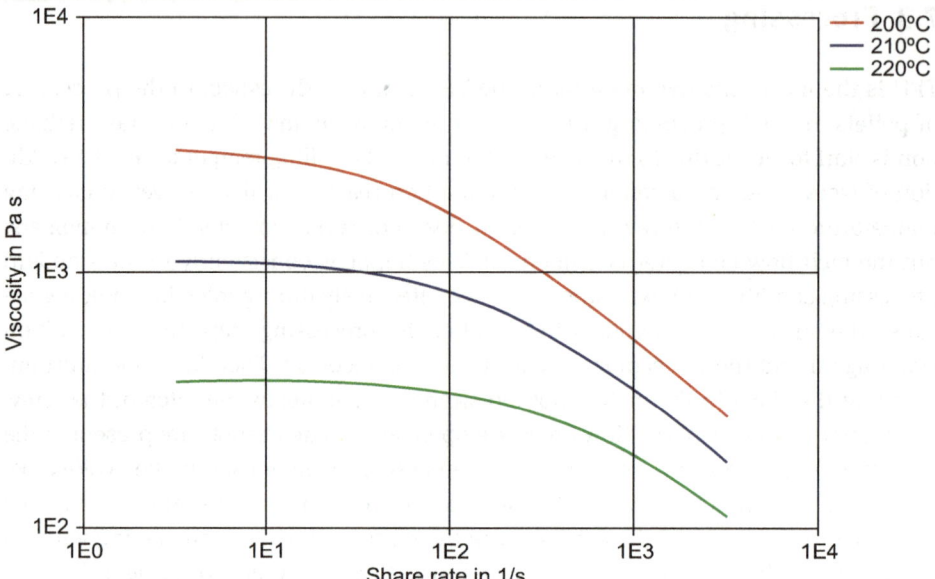

Fig. 7.12: Viscosity as a function of shear rate as measured in a high-pressure capillary viscometer of TPU–ARET of BASF (Shore 97A – 3s) (source: Campus database).

the material and it would not be possible to process the polymer for cable jacketing, tubes, and films on a large scale in order to sell these products.

In the market of extrusion products, customers often ask for the melt flow rate (MFR, see Section 11.6) because they have experience processing PVC or polyolefins and are accustomed to MFR. The information is merely an indication of whether the polymer has a high or low viscosity. But for large volumes, the processor can work with such a value, especially when a consistent material quality is received, and the machine parameters can be held constant.

It was previously mentioned that it is not practical to use a thermoanalysis (DSC) as a processing criterion for TPU because the results are not reliable. TPU changes its morphology during the measurement and its bearing on the melting curve is redundant. At best, the recrystallization behavior of the cooling curve gives a hint about the freezing behavior, and more important is a good knowledge about the processability and choice of a machine parameter.

8 TPC – ester-based TPE is introduced

One specialty group within the thermoplastic elastomer (TPE) family is the thermo-plastic copolyester (TPC) elastomers, which are often used in technical fields where high heat resistance and good processability are needed. One main application area is the automotive sector because the material presents a long lifetime, even under the hood. Close to the engine a suitable temperature resistance is requested. This will be shown a bit later. One invisible application is the use of film. For example, textiles have to be protected against environmental impact and should be sterilized in medical areas. Good processability and high strength make such apparel surfaces durable possible because thin films made from TPC are water protectable and breath-able (Fig. 8.1).

Fig. 8.1: Waterproof textile film for apparel (source: Getty Images).

The crystallizing hard phase is a polyester, mainly polybutylene terephthalate (PBT). This means that the hard segment is made of aromatic structures with very durable building blocks. The flexible soft segment is comparable with those of TPU, namely, polyetherole or polyesterole and are often represented by polytetrahydrofuran or pol-ycaprolactone, and sometimes by polycarbonate dioles, and, the nomenclature of these materials is indicated in the norm ISO 18064 as follows:

https://doi.org/10.1515/9783110739848-008

TPC-ES   Copolyester with polyester soft segment
TPC-ET   Copolyester with polyether soft segment
TPC-EE   Copolyester with polyether–ester soft segment

As with TPU, it is useful to classify according to the soft phase which is still not accounted for in this norm, e.g. a soft segment of polycarbonate diol.

TPC-CE   Copolyester with polycarbonate soft segment

8.1 Manufacturing process

The polymerization of a polyether-TPC is carried out from the monomer terephthalic acid with butanediol to build a PBT hard block and subsequently a polycondensation with a polymer diol to build up the soft segment. It is a typical polycondensation process with water as a by-product (Fig. 8.2). Production in two steps is key for precise reaction control in order to avoid any ester interchange during the polymerization to the desired block structure of a stable PBT and an elastic polymer. The high tendency of PBT to crystallize is a precondition to get a significant phase separation, and good mechanical properties are expected.

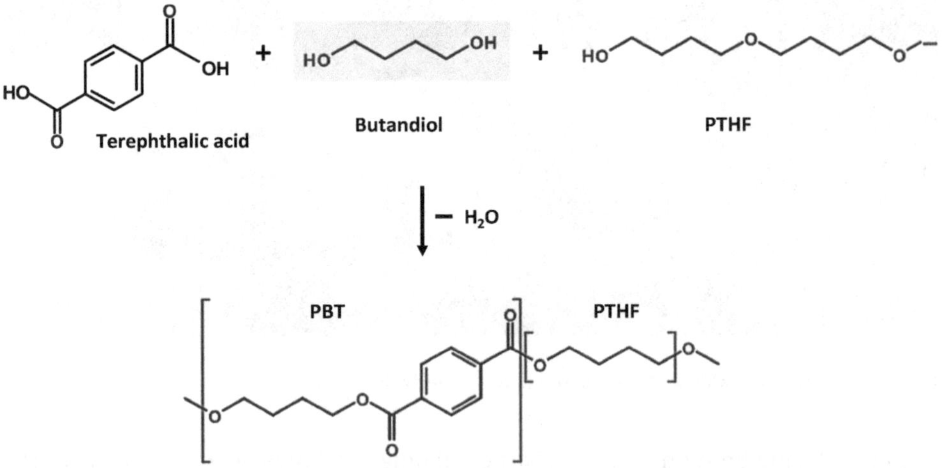

Fig. 8.2: Polycondensation of terephthalic acid with butanediol and polytetrahydrofuran (PTHF).

Similar to Fig. 8.2, the synthetic pathway of a polyester is created with a hard segment out of an aromatic acid and butanediol and a soft segment comprised of a polymeric diol with the same acid. Similar to TPU (Chapter 7), the ratio of hard and soft segments is directly responsible for the shore hardness, and by association the modulus of the material. The manufacturing of TPC takes place in a continuous process in a vessel combination to provide a better temperature control for building the PBT and consequently the soft segment (Fig. 8.3). The final copolymer will be transferred into an extruder in molten form, where homogeneous pellets are subsequently produced.

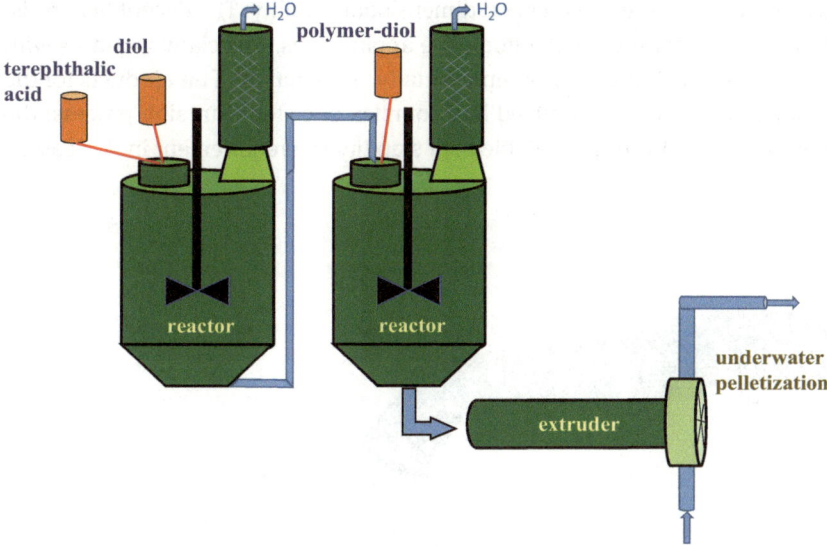

Fig. 8.3: Manufacturing of TPC in a vessel combination.

To some extent, the synthesis is made with dimethyl terephthalate (DMT) and buta-nediol to build up the PBT block by ester exchange under creation of methanol as by-product and subsequently the condensation with a polyol for the soft segment. In the past, terephthalic acid was not easily handled or purified compared to DMT. A newer process and suitable catalysis can do that, could get rid of methanol, and can be made cheaper and more environmentally friendly. Regardless of that, some developments are working on a continuous process in an extruder.

8.2 Properties

The amount of hard segment defines the modulus and the soft segment dictates the level of flexibility. With regard to the soft segment composition, the same properties apply as with TPU, as discussed in Section 7.2. A polyester soft segment is very sus-ceptible to hydrolytic degradation and a polyether soft segment is prone to oxida-tion. An alternative made of polycarbonate diol is a suitable compromise but not a cheap one. On the other hand, a compromise is state of the art to combine an ester with an ether polymer diol. This is visible in the nomenclature by TPC-EE.

The stepwise polycondensation enables a segmented build-up of the block chain, which creates a distinct phase separation. This means that the hard and the soft phases are very pure and not hindered. Therefore, TPC have a good heat resistance and they become solid quickly after processing. These are very important criteria for making molded parts by injection molding. With regard to an extrusion process, the melt stability provides resultant extrudates with good dimensional stability. The decent heat resistance is readily used under the hood automotive applications, especially in places with hot air and oil exposure. Figure 8.4 is a representative illustration of an air ducts for the motor in automobile which are produced in a complex one-step extrusion process: the blow molding extrusion. A precise, reliable melt stability is a requirement in this case.

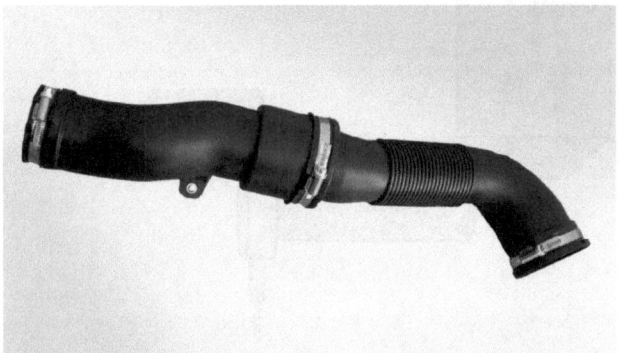

Fig. 8.4: Air ducts for cars under the hood (source: Mocom).

The dynamic thermal behavior over a wide range of temperature can be illustrated via dynamic mechanical analysis (for explanation of that method, see Chap. 11.3) as with other TPE. This measurement provides an idea of the material structure and its utility.

TPC curves (Fig. 8.5) look similar to TPU ones in that the glass transition is located at low temperature and that it appears more distinctly that the grades are softer, which means there is a higher soft phase content. On the right side of the plot, the curves bend at high temperatures, which hints at higher service temperature for TPC compared to previously introduced TPE. It must be taken into account that the softest grades do not display such a high-temperature behavior. In the middle portion of the graph, between 40 and 60°C, small humps are visible. This is indicative of melting of the amorphous parts of PBT segment and they soften much earlier than their crystalline counterparts. Nonetheless, this is irrelevant to the TPC performance and the heat stability persists because the hard phase is regular and pure, enabling a sufficiently high melting point.

Moreover, a peculiarity is visible in that diagram. The soft segments are well separated in this copolymer and tends to crystallize easier at low temperatures than a less pure phase. This can be visualized by the curves in the range from −40 to 0°C, especially

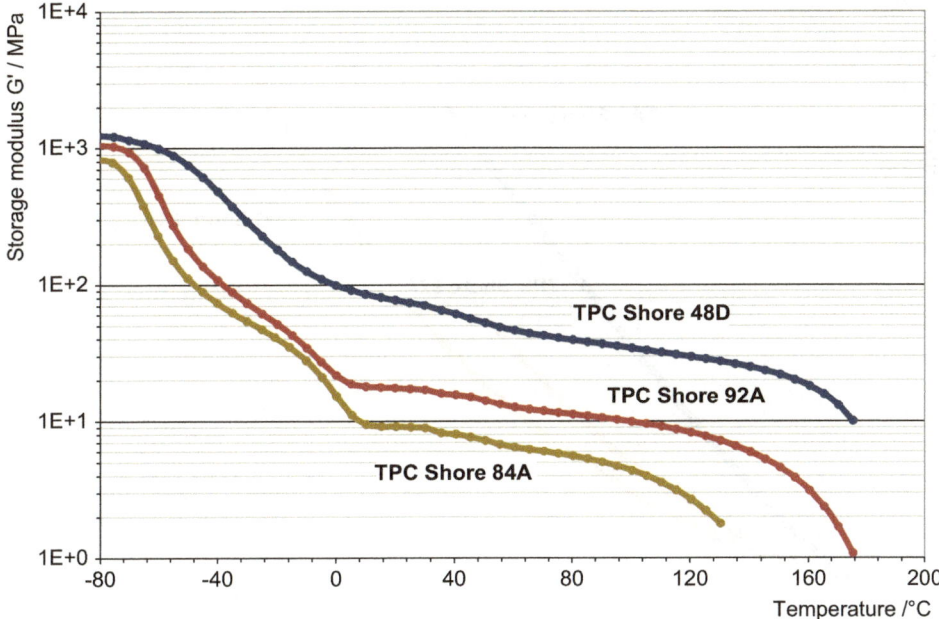

Fig. 8.5: Dynamic mechanical analysis of a TPC-ET from DuPont (Shore 84A – 3s), (Shore 92A – 3s), and (Shore 97A/48D – 3s) on a test specimen in torsion mode at a frequency of 1Hz.

for the soft grades. There is also a change in slope at the end of the glass transition, which reduces the cold flexibility in some cases. Nevertheless, this can be avoided through modification of the soft segment when good cold flexibility is a critical aspect.

The general hardness range of unfilled TPCs from Shore 80A to 80D and the range of mechanical properties for the group are reported by tensile strength, elongation at break (DIN 53504, ISO 37, ASTM D 412), and the abrasion loss (DIN ISO 4649, ASTM D 5963):

Tensile strength	10–60MPa
Elongation at break	300–800%
Abrasion loss	20–100mm^3

Roughly speaking, the property profile of TPC compared to a TPU is very similar. Softer TPU grades are more often used whereas TPC are popular for middle hardnesses like Shore 90A and above. Logically, this middle hardness is where the advantages of TPC are best exhibited.

A deeper look into the elastic properties presents a picture (Fig. 8.6) of an elastomer, also illustrated by the intermittent stress–strain measurement (for explanation of that method, see Chapter 11.4).

In previous graphs of other TPE types, there is typically an initially flat increase followed by a change in the slope. This behavior is not visible here, rather a continuous

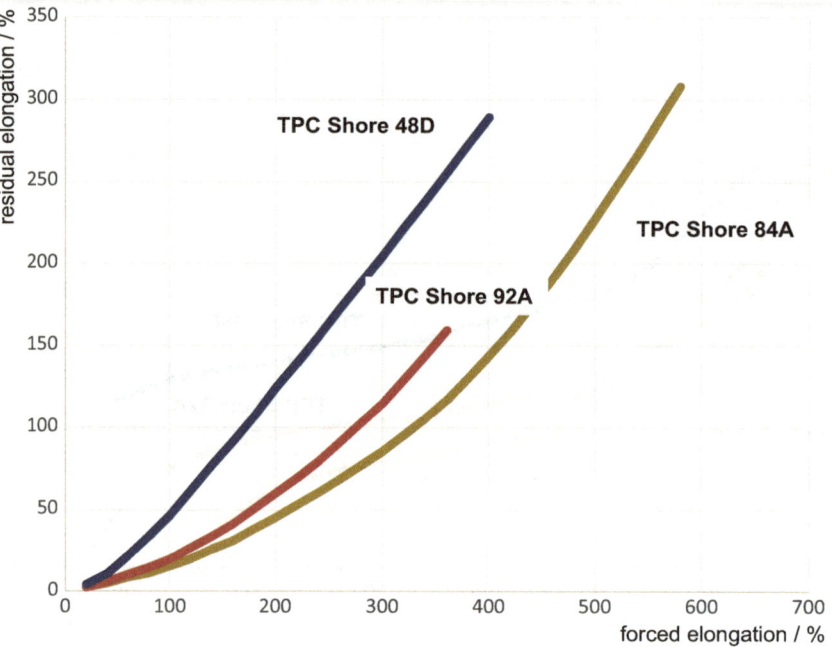

Fig. 8.6: Strain values from the intermittent stress–strain measurement of TPC-ET from DuPont (Shore 84A – 3s), (Shore 92A – 3s), and (Shore 97A/48D – 3s).

increase is observed. The material is still flexible at low elongation, and for application ranges of TPC there is no need for a low hysteresis or a rubber-like behavior. Good heat resistance combined with cold flexibility are the properties predominantly sought. High flexible rubber behavior is seldom asked in the TPE family, and very soft grades are hardly offered.

The tubes in Fig. 8.7 are of a less spectacular appearance but a long-term use under harsh conditions at elevated temperature and dynamic load are requested. In some cases, the contact to certain media plays an important role.

Fig. 8.7: Pressure tubes of TPC (source: Sipol).

8.3 Processing

The good processing behavior of TPC was previously mentioned and is attributed to fast recrystallization and a decent melt stability. In Fig. 8.8, the thermoanalysis (DSC; for explanation of that method, see Section 11.5) of a TPC with hardness of 48D is illustrated.

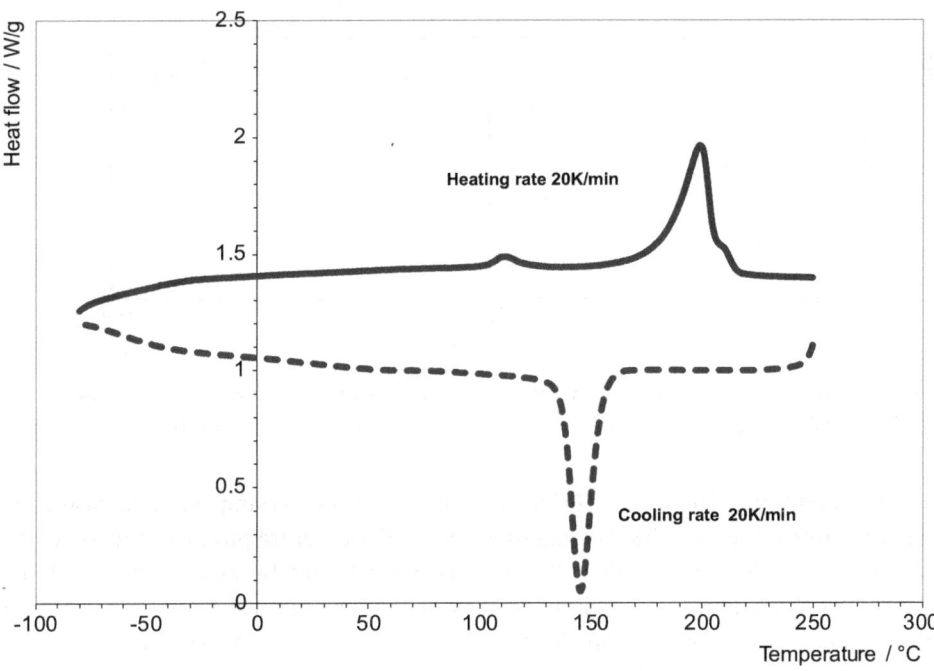

Fig. 8.8: Differential scanning calorimetry of TPC-ET of DuPont (Shore 97A/48D – 3s), at a temperature ramp rate of 20K/min.

The heating curve primarily shows a sharp peak at 200°C, which is characteristic of the melting point of PBT. Similarly, the recrystallization peak in the cooling curve occurs at a relatively high temperature, which means TPC solidifies very early, enabling short cycle times for melt processing. This can be expected because the recrystallization comes back to the baseline very early.

The viscosity curves in Fig. 8.9 predict a generous processing window. This measurement, performed in a high-pressure capillary viscosimeter, is described in more detail in Section 11.7. A plot of the melt viscosities as a function of the shear rate at different temperatures appears close together, and therefore a low sensitivity to temperature variations in the melt during extrusion or injection molding is expected.

The spacing between the curves at these temperature differences is so small that similar viscosities over the shear rate of trial materials can be expected even at

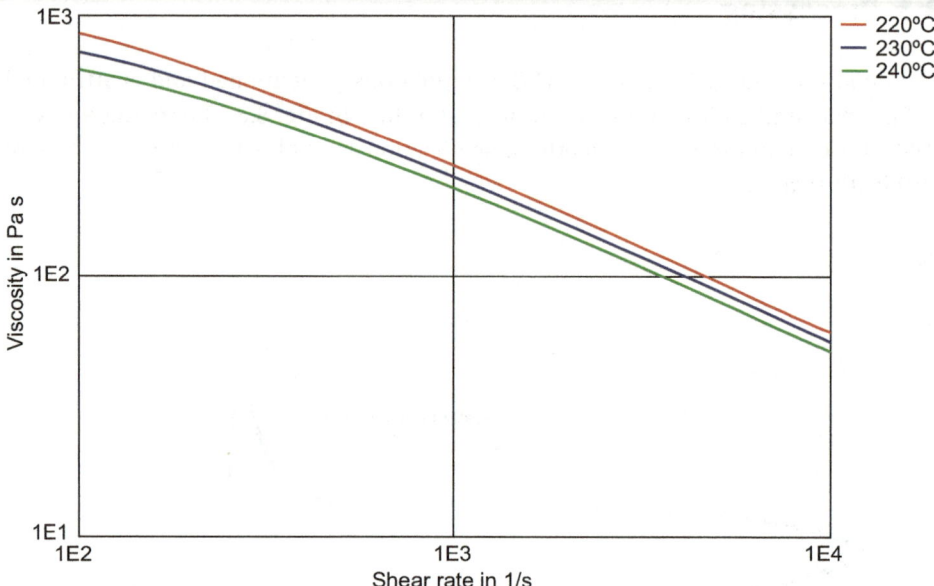

Fig. 8.9: Viscosity as a function of shear rate measured in high-pressure capillary viscosimeter on TPC-ET of DuPont (Shore 97A/48D – 3s), drawn from Campus Plastics database.

various conditions. Therefore, TPC is not sensitive in processing. In most cases, the melt flow rate (MFR) is suitable enough to set the temperature profile of the extrusion machine. For injection molding, the MFR is not relevant because of much higher shear rates. Sometimes it is helpful to look at flow curves, which are available for many grades in the Campus Plastics database (link or reference). This may be useful for calculations and more detailed planning which depends on the processing behavior of a material.

9 TPA – amide-based TPE is introduced

A little bijou in the thermoplastic elastomer (TPE) family is the thermoplastic polyamide (TPA) elastomer group. The prevalent grades in the market are composed of a hard segment of aliphatic amides, mainly polyamide 12 and a polyether soft segment. Regarding the flexible phase, the same variations as the TPC (thermoplastic copolyester) soft phases are possible. Accordingly, these materials are named in the ISO 18064 in the following manner:

TPA-ES   Polyamide with polyester soft segment
TPA-ET   Polyamide with polyether soft segment
TPA-EE   Polyamide with polyether–ester soft segment

The current norm does not acknowledge that aromatic hard segments are also possible but such grades are rarely present in the market. This may be influenced by the large property profile overlap of presently established TPC materials. Figure 9.1 presents an application in the sport shoe area, where high dynamic impact, clear transparency, and a low density are requested. This is an attractive niche among the shoe market, dominated mainly by TPU (thermoplastic urethane), but TPA prefered in the high performance area.

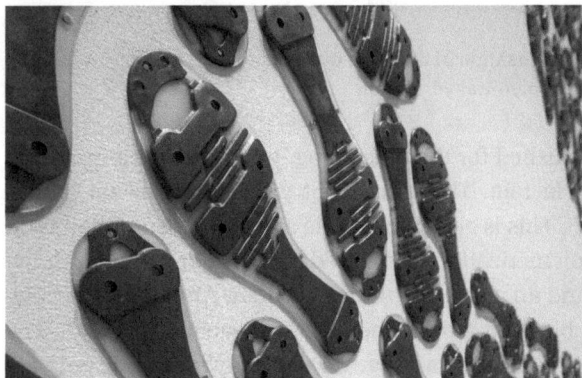

Fig. 9.1: Shoe soles for soccer shoes of TPA (source: Evonic).

The advantages provided by the aliphatic hard segment based on PA12 are the low density combined with a high transparency and light stability of these products. This is a big advantage in outdoor applications. Furthermore, a wide processing window offers even further benefit. All these properties can be seen in the illustrated plates for high-grade soccer shoes.

https://doi.org/10.1515/9783110739848-009

9.1 Manufacturing process

Older processes are based on a condensation reaction of diamine and diacid in a vessel producing H_2O which must be removed. Then, amide blocks with acid end groups are built up, which are subsequently reacted with a polymeric diol, in the following example (Fig. 9.2), a polyether from tetrahydrofuran (PTHF or PTMEG), again via condensation of water. The rigid polyamide segment will be created from amine and acid, and the soft segment comes from the connection of a polymer diol.

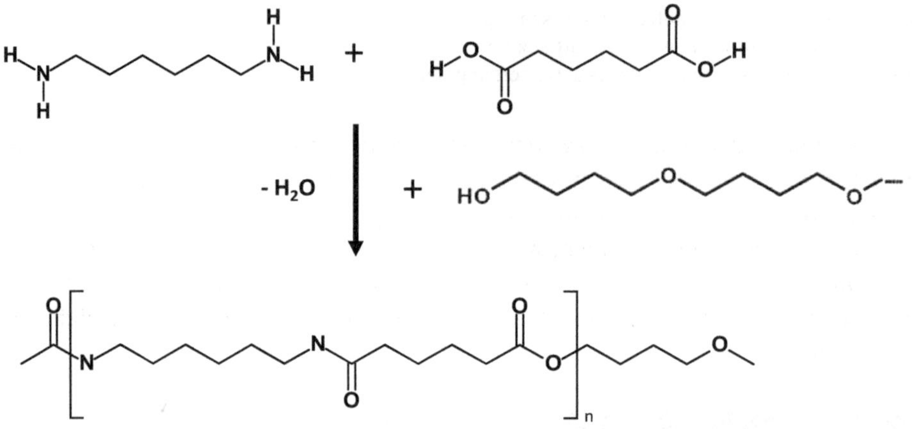

Fig. 9.2: Manufacturing of TPA by polycondensation of diamine, diacid, and diol; here hexamethylene diamine, adipic acid, and polytetrahydrofuran.

Only the aliphatic-based TPA established for the time being, which are built up by the ring-opening polymerization of a lactam. The incumbent grade is based on laurinlactam (for PA12 blocks) and PTHF. This is easier to handle and the consumption is on a good scale. The ring-opening polymerization is initiated by alkali or alkali earth hydroxides and a polyamide block and an acid end group is created. After that, the polyether diol is connected to the PA block in a polycondensation reaction (Fig. 9.3). The reaction is carried out in a vessel under inert gas (Fig. 9.4). Subsequently, the polymer

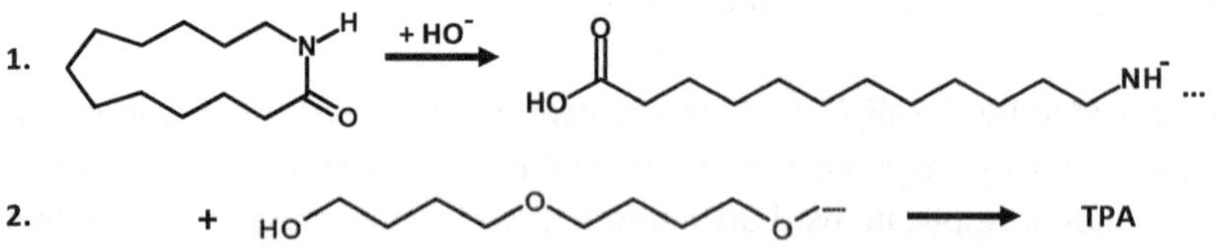

Fig. 9.3: Reaction of TPA with ring opening of laurin-lactam and the following condensation of polytetrahydrofuran.

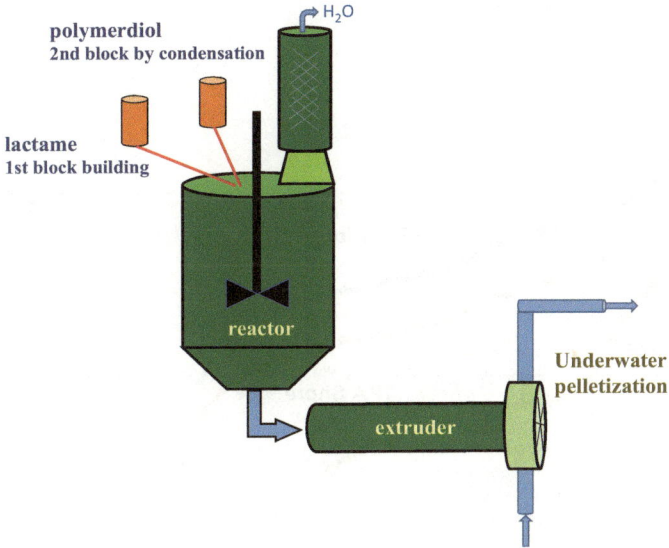

Fig. 9.4: Manufacturing of TPA in a vessel from lactam and polymer diol.

melt will be transferred into an extruder and finalized via an usual granulation process, cutting rods or underwater pelletization.

9.2 Properties

Bearing in mind that aromatic polyether–amides are scarcely visible in the market, here the relevant TPA based on PA12 and polyether soft phase will be discussed. They are heat resistant, transparent, and easily processed. Their property profile contrasts with the other TPE considerably, but they have a low density, similar to polyolefins few above 1g/cm^3. Like TPC, TPA also have a good phase separation of their building blocks; the hard segments are stable and the soft segments are quite pure.

Again, the dynamic mechanical analysis (DMA) is displayed, and Fig. 9.5 shows three TPA with different hardnesses (explanation of that method see Section 11.3).

It is clearly seen that the material transitions from stiff to elastic status happens very early at low temperature. This causes a good cold flexibility and used at very low temperature. It can be expected that these materials are not brittle at –40°C. It should be taken into account that the very homogeneous soft phase (PTHF shown here) tends to crystallize and the cold flexibility is visibly reduced. In the previous chapter on TPC (Section 8.2), the same behavior of humps in the curves is seen in DMA. If desired, the polyol may be easily modified to eliminate this effect. With exception of the softest sample, the TPA present a plateau over a large temperature range and according to the DMA curve, these materials can be used up to temperatures of 120°C.

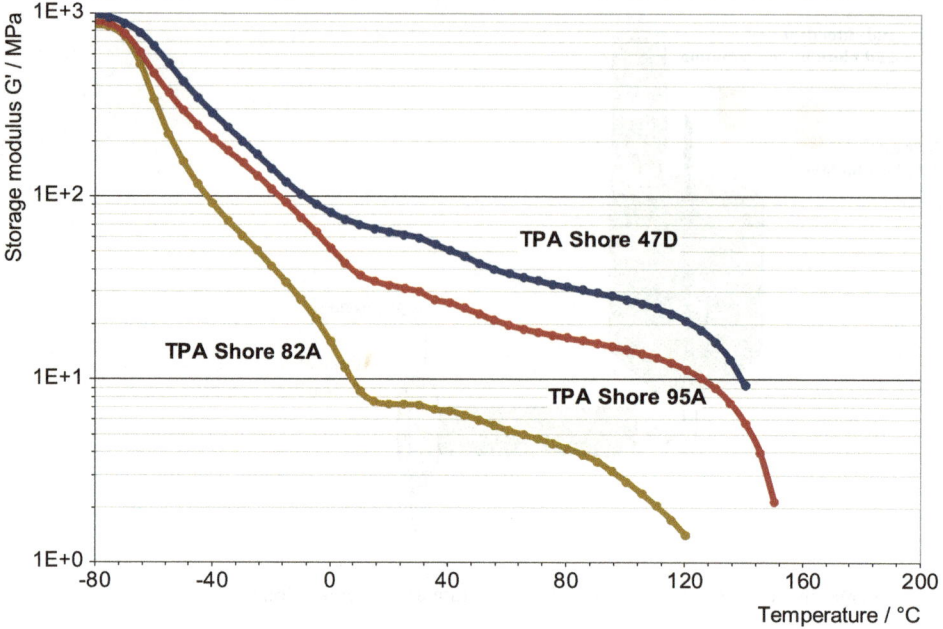

Fig. 9.5: Dynamic mechanical analysis of a TPA-ET from Arkema (Shore 82A – 3s), (Shore 95A – 3s), and (Shore 97A/47D – 3s) on a test specimen in torsion mode at a frequency of 1Hz.

The following values evaluated from TPA in a range of hardness Shore 70A until 70D describe the tensile strength (DIN 53504, ISO 37, ASTM D 412) and abrasion loss measurements (DIN ISO 4649, ASTM D5963) for most of the available products:

Tensile strength | 30–60MPa
Elongation of break | 200–800%
Abrasion loss | 40–130mm^3

Soft grades play a less prominent role in the market because their mechanical properties are not better than those of cheaper materials. The advantages of TPA begin to shine at middle Shore A values. Otherwise compared to TPC and TPU, the structure of the soft segment is responsible for the resistance to environmental effects. Polyethers are sensitive to oxidation and UV irradiation, and polyesters must be hydrolytically stabilized. These aspects had been discussed in the chapter of TPU (see Section 7.2) and can be transferred to the circumstances of TPA soft segments. Overall, TPA are stable in oil and grease due to their polarity.

Pictured here, the intermittent stress–strain measurement (Fig. 9.6) provides an idea about its elastic behavior and deformation at high extension (explanation of that method see Section 11.4).

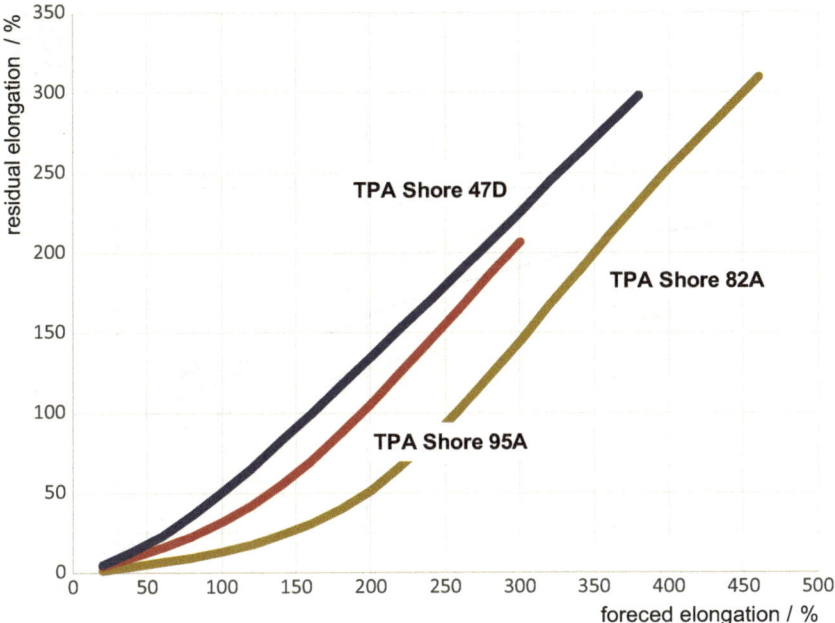

Fig. 9.6: Strain values from the intermittent stress–strain measurement of TPA-ET from Arkema (Shore 82A – 3s), (Shore 95A – 3s), and (Shore 97A/47D – 3s).

From the beginning of the measurement until a strain of about 100%, the distinct difference in material stiffness is clearly visible, as described. Whereas the material with Shore 82A behaves highly elastically, the deformation of the other samples increases with higher modulus. It can be assumed that the amorphous hard segment is less stable under high elongation. Normally, TPA are not used in applications that need high elongation. Most likely the wear resistance and high ductility combined with low density and transparency are the key combinations driving the selection of a TPA.

9.3 Processing

As mentioned previously, the viscosity curves at different temperature from high-pressure capillary viscosimeter (explanation of that method see Section 11.7) give an impression of the ideal processing window of the tested material. Considering the wide temperature range of Fig. 9.7, the curves are very close together. The material is not highly sensitive to the melt processing temperature, which is an advantage for extrusion, as an example, in order to make large diameter hoses accurately even if the die suffers from inhomogeneous heating.

In many cases, the melt flow index (melt flow rate) is sufficient to set temperature profiles on the machines but sometimes more information is desired, and it is

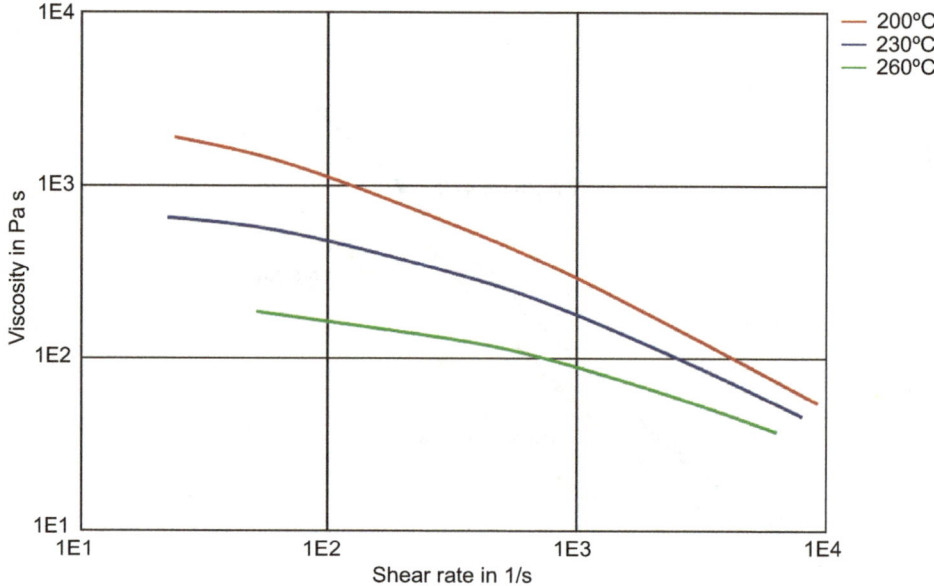

Fig. 9.7: Viscosity as a function of shear rate measured in high-pressure capillary viscosimeter on TPA-ET from Arkema (Shore 95A – 3s), drawn from Campus Plastics database.

recommended to seek these flow curves from the CAMPUS database, where TPA is represented as well. This is important when calculations must be done, or more detailed processing data are needed. As mentioned earlier, the good melt stability of the TPA translates into a well-behaved dimensional stability. The particular property profile is illustrated in Fig. 9.8. The skills of a material for spectacle-frames are a low density, break resilience, high transparency and heat resistance.

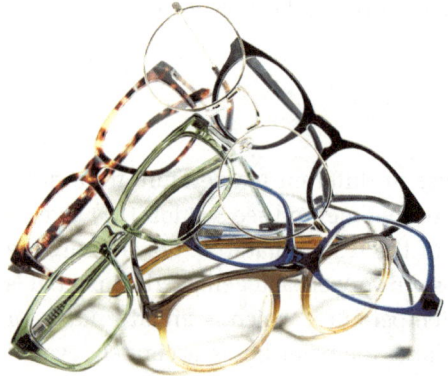

Fig. 9.8: Transparent spectacle-frames (source: filipw/iStock/Getty Images).

In preparing to process a TPA, a thermoanalysis measurement (DSC; for explanation of that method), see Section 11.5) is not very important. Nevertheless, Fig. 9.9 shows both a clear peak in the melting area and a distinct peak upon cooling indicating a fast crystallization, that is, solidification. This reinforces that a reasonable cycle time for melt processing can be expected.

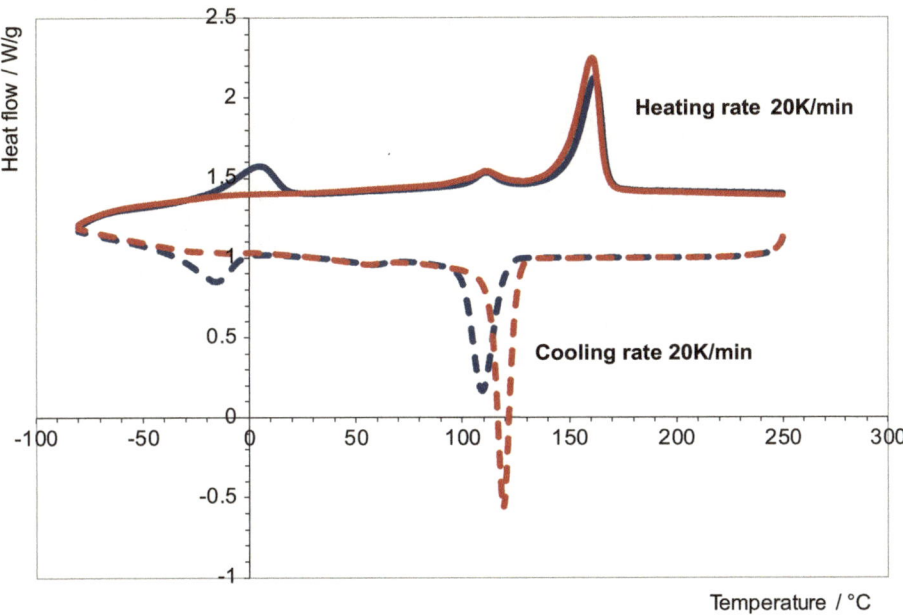

Fig. 9.9: DSC of TPA-ET of Arkema (Shore 95A – 3s) – blue line – and (Shore 47D – 3s) – red line, at a temperature ramp rate of 20K/min.

That said, not all materials of one family behave in a same manner. A look at the DSC above, the soft TPA grade presents a crystallization of the soft phase below room temperature (blue line in Fig. 9.9). The reason is a quite clear phase separation related to the building blocks of the synthesis, which means a pure hard and soft segment. Here, we see the crystallization of the linear soft segment, influenced by its molecular weight and the structure. This agrees with the hump in the related DMA diagram (Fig. 9.5). As mentioned, this effect can be easily eliminated by modification of the soft phase when a more restrictive flexibility in the cold is required.

10 Renewable materials

Bioplastics, biobased plastics, and biodegradable plastics are on everyone's lips in the polymer world. People discuss about environmental pollution, which arises from inappropriate handling of these reusable materials. Here, it is not the intention to discuss how to proceed with a recycling process or the proper combustion of waste material. Landfill is the least desirable route to dispose of plastic materials because of their poor and very lengthy degradation in the environment. It is barely conceivable outside of a controlled industrial process.

The manufacturing of raw materials from plants as an alternative to a petrochemical source is established since a long time. These are called biobased or renewable raw materials. Even the critique of sourcing from the food chain will not stop the trend and prevent development in this route. From fermentation technology research, there are promising results that find a way of producing chemicals for polymer-building elements from food and biological pollution in the future.

In relation to these activities of finding biobased monomers for plastic materials, thermoplastic elastomers (TPE) are also involved. Building blocks such as those from the oil or gas supply chain already exist. One of the most effective sources is sugar to receive carbon in C_2, C_3, and C_4 units from a fermentation process. The most exposed material in that case is ethanol for its use in biofuel for cars. The chemically reduced version of ethanol, the ethylene, is used as a monomer for thermoplastic polyolefin and ethylene–propylene–diene–rubber, the relevant component in thermoplastic vulcanizate. The base material is glucose, which is extracted from starch-containing plants like sugar beet, corn, and grain, but sugarcane is still the most effective source of such raw materials, getting diols from fermentation. Even though the acid itself, like malonic or succinic acid, will not be used in the polymerization directly, diols can be made by reduction reaction and can be used in synthesis with acids to create a polyester diol. The diols are the basis for polyether diols as well (Fig. 10.1). Polyester and polyether (telechels) are building elements for the soft phases of TPU (thermoplastic urethane), TPC (thermoplastic copolyester), and TPA (thermoplastic polyamide).

polyether-diol (here: polytetrahydrofuran)

polyester-diol (here: polybutylen-adipat)

succinic acid → butanediol

Fig. 10.1: Reduction reaction of succinic acid into butanediol and following polymerization to polytetrahydrofuran or polycondensation with adipic acid to polyester diol.

https://doi.org/10.1515/9783110739848-010

This is an example for raw materials out of four carbon atoms. Likewise, building blocks with three carbon atoms are made to get polyester or polyether (Fig. 10.2).

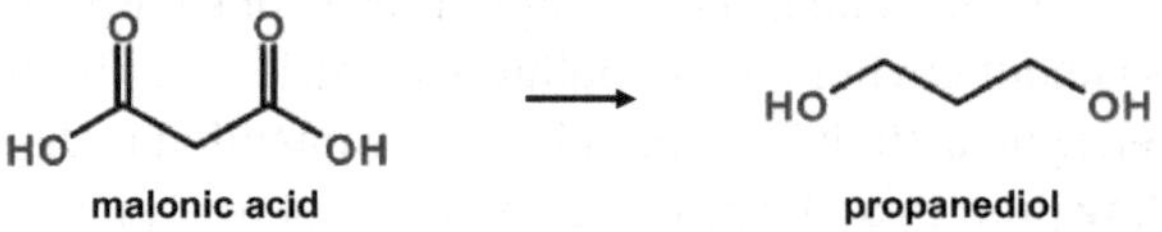

malonic acid → propanediol

Fig. 10.2: Reduction reaction of malonic acid into propanediol.

There is a high interest in biobased TPE in the market, and suppliers offer several solutions but the business remains relatively low. Nevertheless, the trend will continue, and the consumption in manufacturing and the product quality also improve more and more. When these building blocks are available, the TPE can be synthesized in the same equipment and process. Furthermore, functionalization of these moieties provides building blocks even for hard segments like terephthalic acid in TPC or specific amines in TPA.

From castor oil, sebacic acid (C_{10}) can be made which is used as a polyester for the soft phase. The same basis is used to make undecane acid (C_{11}) to transfer into an amine for the polymerization of polyamide 11. It should be considered that it is not a simple task to harvest castor oil, because the plant very prickly grows only in a special climate of a certain region. As a positive result, food chain is not touched.

Of additional interest is lysin, a substance sourced from a starch biomass in a fermentation process (Fig. 10.3).

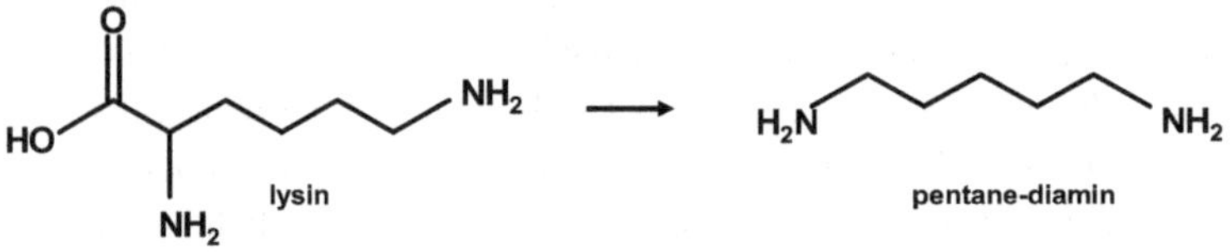

lysin → pentane-diamin

Fig. 10.3: Pentane diamine from lysin.

The residual pentane-diamine (C_5) has the potential to build the hard segment of a TPA or, with further functionalization from diamine to diisocyanate, the hard phase for TPU.

A new technology can make polycarbonate diols from CO_2. This polyol can be used as is in the soft phase of TPE like TPU, TPC, and TPA or in combination with other polyester and polyether. This consumption of carbon dioxide is already introduced in the plastics world.

Even when it is not the focus of this chapter, it is worthwhile to mention the recyclability of these thermoplastic materials, which is also an important element in the circular economy. TPE are thermoplastic and can be molten anyway. In-house recycling and its continual improvement have been state of the art in the plastic industry for many years. Reasonable recycling of used articles is always possible depending on the logistic effort, cleaning of the articles, and the separation of material combinations into single components.

Additionally, there are ongoing developments to collect the used plastic parts, degrade them in a pyrolysis process, and ultimately produce monomers for a new polymer. Pyrolysis is a burning process in a completely close chamber system and does not exhaust any gases. This sounds relatively simple, but the detailed process will still require some effort.

11 Description of testing procedure

11.1 Shore hardness by ISO 48-4 (formerly ISO 7619-1)

The most popular method to classify thermoplastic elastomers is done through the testing of hardness on the shore scale, which is described in the standard ISO 48-4 (formerly ISO 7619-1). A test probe of defined shape is applied with a defined load onto the flat area of a test specimen, and the resistance gives the hardness value. Three different probe geometries are used to cover the wide range of elastomers. The harder the test specimen is, the sharper the probe tip must be. Fig 11.1, extracted from the named standard, shows the different geometries.

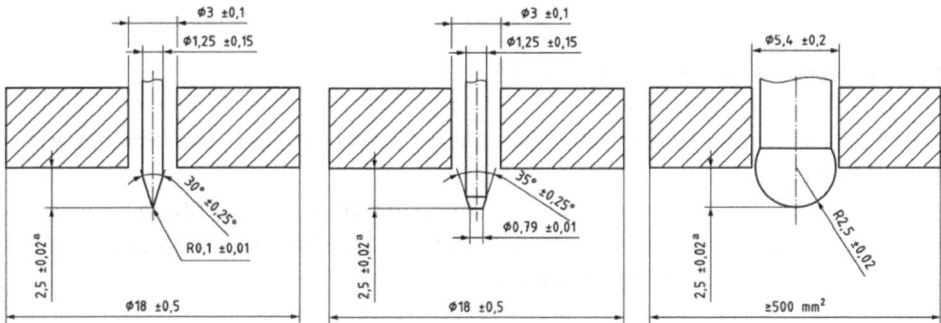

Fig. 11.1: Measurement pin for Shore hardness from ISO 48-4.

On the left side, a sharp needle is illustrated for use on hard specimens, which are classified in Shore D. In the middle is the flat probe for softer grades in the range of Shore A. The values for very soft materials are measured by Shore A0 (or Shore 00) with a needle point like a ball. It is reasonable to expect numbers in a range above 0 and below 100, preferentially comfortably within middle of this range. Among these three classifications, there are overlaps, although some vendors provide two values. These appear more often on thermoplastic copolyester and thermoplastic polyamide data sheets, where numbers of Shore A and D are indicated.

One important criterion is the dwell time until the measurement is recorded. The standard recommends both 3 and 15s. Measurements with both dwell times are particularly important for very soft grades because the test probe continues to penetrate during the measurement.

All values of presented material samples fall within Shore hardness A or D and are recorded after 3s penetration time. This is the most used method for thermoplastic elastomers (TPE) in technical applications. As mentioned earlier, there is no mathematical relation between Shore A and D. Fig 11.2 illustrates the most recommended area for both hardness, denoting that Shore A should be used until 95A and thereafter, Shore D

https://doi.org/10.1515/9783110739848-011

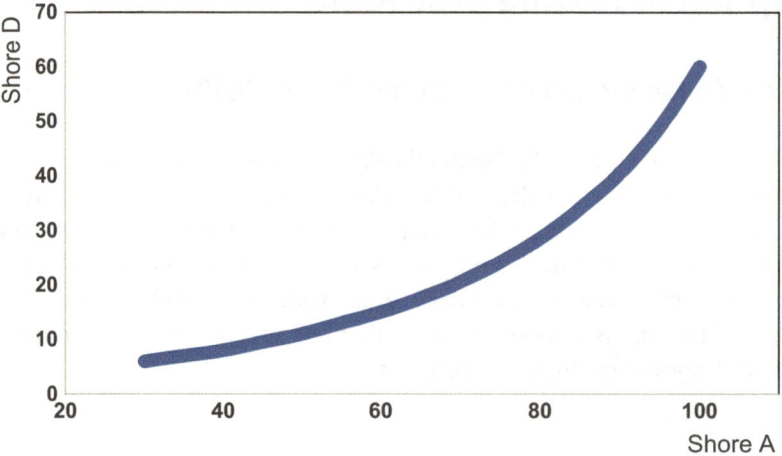

Fig. 11.2: Relation between Shore A and Shore D.

as of 50D. For measurements with values that fall into both ranges, both Shore A and D should be reported.

Below Shore 40A values are less reliable, and an additional Shore 0 measurement is recommended. Such a material behaves more like a gel and even the flat needle does not stop during penetration.

11.2 Compression set by ISO 815-1

For specific applications it is important to know how an elastomer changes its shape under a static load and recovers after a defined time. This can be reported by the measurement of compression set as described in the standard DIN ISO 815, where different conditions can be found depending on the request of certain application. A cylindrical test specimen of defined shape is clamped between steel plates at a defined height. The entire apparatus is placed into an oven at 23, 70, 100, and, possibly, 150°C depending on the test requirement and nature of the material. The samples are released according to their procedure. In method A, the plates are opened immediately, and the material recovers at room temperature. Method B keeps the sample under pressure at room temperature, and in method C, the material is released immediately but recovers at test temperature.

The compression set is an indication of the heat resistance of a polymer, because under load and temperature it deforms faster and changes its shape. In the area of seals, gaskets, rollers, or buffers, this is a helpful measurement to judge material suitability.

11.3 Dynamic mechanical analysis by ISO 6721

All measurements of dynamic mechanical analysis, or dynamic mechanical thermoanalysis are done under the direction of EN ISO 6721 with the same equipment under the same conditions. The test specimen are straight, 12mm broad, and 2mm thick, cut from an injection molded plaque. The test is carried out under dynamic torsion with a 1Hz oscillation frequency on a test bar which is prestressed by 20%. During this procedure, the specimen is kept in the machine at one conditioning temperature until a constant modulus value is reached before the measurement begins. The deflection of approximately 0.1% is low enough that a linear viscoelastic behavior of the polymer is expected, implying that the modulus is constant under that deformation. In the next step, the temperature is increased in increments of 5°C and for every subsequent measurement. Thus, a modulus value is detected for each step while the test bar is in a conditioned status. The equipment (Fig. 11.3) is constructed so that the sample is stressed from the clamp at the bottom and the sensor is installed on the top. This ensures that the modulus is measured along the whole specimen.

Fig. 11.3: Model of dynamic mechanical analysis (DMA).

The deformation, γ follows the sine function (ω = angular velocity, t = time)

$$\gamma(t) = \gamma_0 \sin(\omega t)$$

The torsion τ of the test piece depends on the function of a specific phase deviation δ

$$\tau(t) = \tau_0 \sin(\omega t + \delta)$$

Depending on the phase deviation, the torsion is divided by the storage modulus G' and the loss modulus G''

$$\tau(t) = \gamma_0 [G' \sin(\omega t) + G'' \cos(\omega t)]$$

G' mirrors the elastic part of the material, whereas G'' represents the viscous one. The storage modulus represents the energy returned to the polymer, making the TPE elastic; the energy of the loss modulus dissipates by another pathway in the material. It can be transformed into heat, for example. The quotient of G'' and G' is the deviation angle $\tan \delta$

$$\tan \delta = G'' / G'.$$

Logically, when this ratio exceeds a value of 1, the TPE starts to crossover from a solid into a liquid status.

Here, the modulus was discussed for a torsion measurement and is designated G, indicating a shear modulus. From a tensile measurement, the elastic modulus E (also known as Young's modulus) is given, which should be characteristically similar. The factor between both is 2.5 to 3 times of G, depending on the Poisson value μ (lateral contraction coefficient) of the single TPE material, which is in the area of 0.4–0.5. This indicates the following equation between the elastic and the torsion modulus:

$$G = E / 2(1+\mu)$$

Even here, it is important to evaluate the modulus for the same sample at a very low strain in the linear viscoelastic regime and without any residual deformation.

11.4 Intermittent stress–strain measurement

In the current model, a test specimen with a shape as in Fig. 11.4 of type S2 (DIN 53504) or type 2 (ISO 37) is placed into tensile test equipment and prestressed by a force of 0.2MPa until an elongation of 20% is reached under ambient conditions. The measurement starts with a speed of 50mm/min for an additional 20% strain, and subsequent relaxation until the initial 20% is recovered at the same speed.

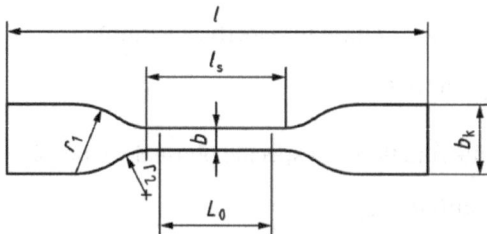

Fig. 11.4: Test specimen type 2 from ISO 37.

This procedure is repeated in 20% strain increments for every step. This continues until a desired elongation (e.g., 600%) or until the test bar breaks.

The width used for the respective test specimen is stated in the standard.

The values of the applied strain and the related residual values after relaxation are assembled in an additional diagram. For a more intensive study on that subject the literature of N. Vennemann is recommended.

11.5 Dynamic thermoanalysis by ISO 11357

In a dynamic scanning calorimetry (DSC), a few grams of a well-known reference material and a test sample are put into little chambers and heated under controlled and isolated conditions. The measurement starts at very low temperature (e.g., −80°C) to see the glass transition of the flexible phase and it ends when the test material is fully molten. The difference in the amount of heat between both samples creates a curve where the phase transitions of the test sample are visible in a plot of temperature versus change in enthalpy. This is the basis of differential calorimetry.

During the initial heating from the frozen state, first, the flexible soft segments of a TPE transitions from a glassy behavior to an elastomeric one. When the material starts to melt, the next transition, or heat flow, is detectable and the amount of heat is represented by the area under the curve (enthalpy). The recrystallization or solidification can be upon cooling at the same controlled speed. The residual lines are the heating and the cooling curve.

11.6 Melt flow rate by ISO 1133

The melt flow rate (MFR) or melt volume rate is a parameter used to roughly estimate the melt flowability of a polymer. Often the expression MFI for melt flow index is still used. The procedure is published in the standard ISO 1133-1 (Fig. 11.5).

Pellets of the sample are loaded into a vertical cylinder, pressed by a stamp, and heated up. Below the cylinder lies a funnel which feeds into a die with defined bore. The measurement proceeds at a defined load and temperature, and the final value of MFR is given by the amount of melt (in grams) which has exited the chamber in the span of 10min. Therefore, a highly viscous material has a low index, and conversely, a low viscosity material has a high MFR.

All parameters, like load on the stamp, temperature, and die geometry, can be varied and nominal conditions are tabulated in the named standard. They should be chosen in that way that the MFR is not too small (<1) and not too big (>100). Exceptions can be made when the customer desires a value at special temperatures to compare that sample with other familiar materials. MFR provides a value of viscosity at a defined

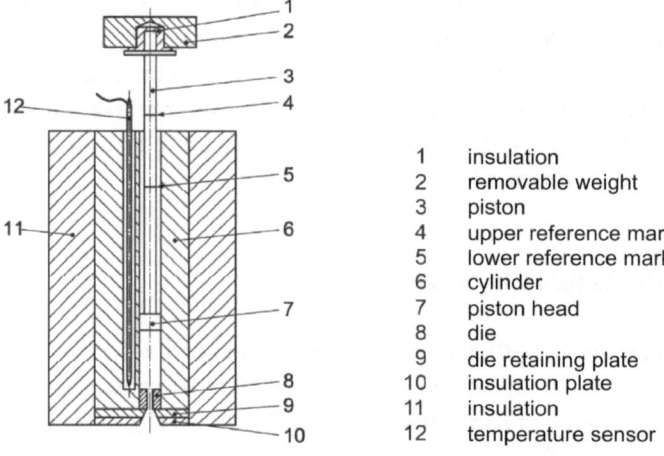

1	insulation
2	removable weight
3	piston
4	upper reference mark
5	lower reference mark
6	cylinder
7	piston head
8	die
9	die retaining plate
10	insulation plate
11	insulation
12	temperature sensor

Fig. 11.5: Model of measuring equipment for melt flow index from ISO 1133.

shear rate, which is below the conditions of an extrusion and way below the conditions of an injection molding process.

11.7 Viscosity curves by ISO 11443

Continuing on the concept of the melt flow rate, the viscosity curve is a visualization of the relation between the viscosity η of a polymer melt and the shear load during processing. As stated by Newton's law, a relation between shear stress σ, shear rate $\mathring{y}$, and viscosity η can be stated as such:

$$\eta = \frac{\sigma}{\mathring{y}}$$

The shear stress comes from the difference of pressure before and after the die and the shear rate is calculated from the volume flow of the polymer melt.

Instead of measuring at only one, low pressure shear rate, the high-pressure capillary viscometer can measure all the parameters that are needed for a complete viscosity curve. The method of that procedure is described in the standard ISO 11443 (Fig. 11.6).

The pellet sample will fit into the temperature-controlled barrel, melted, and the control of the piston's force is able to push the melt over a wide range of shear stress through the die. Together with a variation of dies, the viscosity of the melt can be determined until high shear rates cover all the conditions of used processing methods.

The dependence of the viscosity on the shear rate is not linear for TPE and η decreases faster with increasing shear. This is called structure viscosity, as illustrated in the curves of Fig. 11.7.

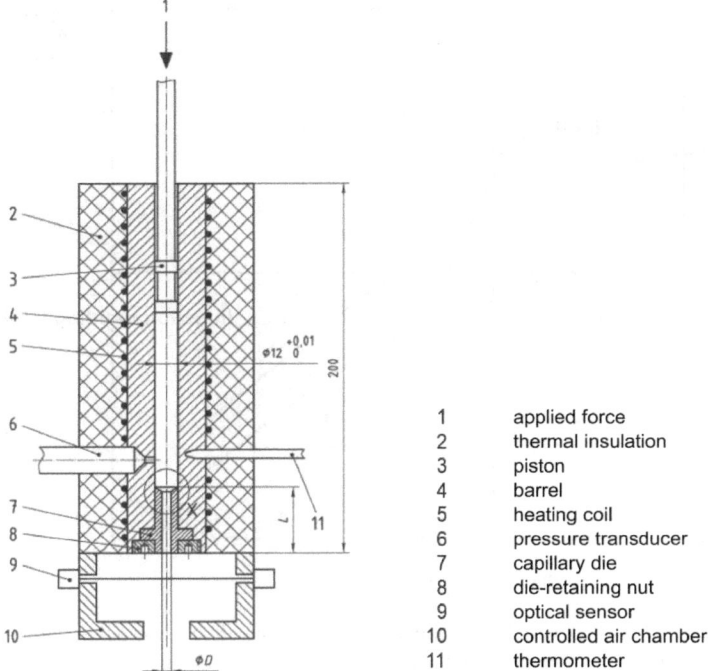

1	applied force
2	thermal insulation
3	piston
4	barrel
5	heating coil
6	pressure transducer
7	capillary die
8	die-retaining nut
9	optical sensor
10	controlled air chamber
11	thermometer

Fig. 11.6: Model of measuring equipment of high-pressure capillary viscometer from ISO 11443.

This diagram illustrates exemplary flow curves of a TPU at three different temperatures (see also Section 7.3) and a rough distinction of ranges of expected shear rates depending on the melt process. The MFR gives only one value and this is below the level of an extrusion process and far away from the injection molding condition. Values from a capillary viscometer provide much more information about the processing behavior of a polymeric material like TPE.

In case of detailed calculation of melt flow data, some corrections of the detected values must be done. For example, there are local deviations when the melt enters into the die and passes through an orifice. In this instance, a different flow characteristic than that described by Newton's law will apply. Normally some of these corrections are installed in the software of a modern viscosimeter. More detailed background concerning the rheology of polymer melt can be studied in the publications of W.-M. Kulicke (*Fließverhalten von Stoffen und Stoffgemischen*) or L.E. Nielsen (*Polymer Rheology*).

11.8 Spiral flow

The community of TPS (styrene block copolymers) and thermoplastic vulcanizate often use a spiral flow mold for the testing in injection molding process. This is very close to

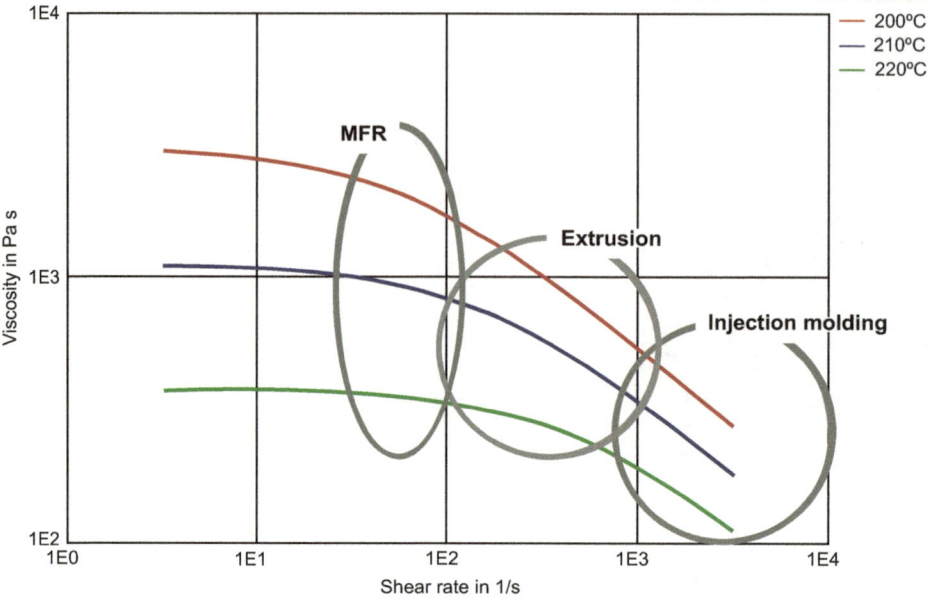

Fig. 11.7: Viscosity curves of a TPU–ARET with illustration of the shear area of MFR, extrusion, and injection molding, drawn from Campus Plastics database.

the practice and a valuable information for the molder. The shape of these pieces is different, but they have the same target. In the channel, marks are punched in to read afterward the length of the injected piece. The longer it is, the lower the viscosity of the melt has been. The geometry of the spiral often relates to the circumstances of the used machines. In many cases, a spherical shape is used and in Fig. 11.8, a rectangular one is illustrated. The sprue is not cut here.

Fig. 11.8: Spiral flow from a TPS (source: Allod).

On the test piece, the marks of every section can be seen easily. In the same procedure, the processing conditions of a sample will be worked out for a best recommendation for the customer.

12 Epilogue

This book provides a relatively quick and yet a deeper view into the diversity of thermoplastic elastomers (TPE): how they behave, how they perform, and how they can be understood. The distinction from classic rubber elastomers helps in the discussion of when to use a TPE and when to select a rubber. The consistent recommendation is that it is important to contact the supplier for a direct consultation from the manufacturer to achieve the desired technical solution with elastomeric materials. The breadth of the performance range of all TPE cannot be collected in one format. Thus, the comparison of properties among all TPE in the family is not genuinely possible.

At this point, it should be mentioned that no market analysis will be discussed in that book, because after a short period of time the results are nearly outdated. In respect to a study in the year 2018 from Freedonia, the following market share of TPE in the world can be stated: entire TPO (thermoplastic polyolefin) 42%, TPS (styrene block copolymers) 36%, TPU (thermoplastic urethane) 9%, and TPC (thermoplastic copolyester) 3%. Thermoplastic polyamide are listed under others (3%). Roughly seen, these number do not vary so much but the biggest unsure fact is the differentiation of TPE to other plastic materials. A clear cut toward hot melts, coating solutions, or even impact-modified engineering plastics is not seriously possible.

Standards help in orientation, creating a consistent direction. The TPE are described in the DIN EN ISO 18064 and classified but their development in the market continues. One example is TPO, which is presented as an olefinic compound, but not as a copolymer that has existed for a long time. To separate TPO blends from an impact-modified polyolefin cannot be made anyway. Additionally, the intention of a change in the standard is naming the real situation that TPS and often TPO copolymers are compounds in technical application. This is close to a correction and a new draft is published. Even distinguishing between a TPE and an impact-modified plastic material is difficult. The norm does not explain this. Moreover, it should be known that most technical applications of TPS are driven by compounds. A proposal for updating these aspects in the DIN EN ISO 18064 is already published as Draft International Standard (DIS).

It is time to officially install TPE as its own product family and some activities in this direction have already started. One part is the open TPE Forum (www.TPE-forum), a network of technical experts from producers, processors, universities, and associations founded in 2016 in Germany. From this group, which is not an organ, teams have been created to work on standards and guidelines, as well as finding opportunities to spread education about TPE. The official contact of the Forum is the WDK (Wirtschaftsverband der Deutschen Kautschukindustrie), which is a formal incorporation of the TPE into the world of elastomers. The plastics database "Campus

https://doi.org/10.1515/9783110739848-012

Plastics" for TPE properties is also a key point. In the German association of engineers (VDI) members of the TPE forum created a guideline (VDI 2020), which recommend the most suitable methods for characterizing the melt of the TPE in particular. A work group in the German Standard Organization DIN is installed for TPE.

The brief *Glance at Thermoplastic Elastomers* contributes to the intention to delve into the TPE world and to better understand and handle the versatility of this product family. Hopefully, it is a small step in elevating TPE as an own established polymer class among all plastic materials.

Literature

For further studies on TPE, the following literature may be suitable

G. Holden, et.al.: Thermoplastic Elastomers[1], Hanser Verlag 2004
J. Drobny: Handbook of Thermoplastic Elastomers[2], Elsevier 2014
T. Dolansky, M. Gehringer, H. Neumeier: TPE Fibel[3], Dr. Gupta Verlag 2007
Four times a year, *TPE Magazine* of Dr. Gupta Verlag[4]
Brochures and data sheets of TPE supplier
Database Campus® Plastics via Internet [www.CAMPUSplastics.com]
DIN EN ISO 18064 for nomenclature of TPE
ISO 14910 for TPC
ISO 16365 for TPU (new project in DIN)
ISO 48-4 for measuring the Shore-hardness, formerly ISO DIN EN ISO 7619-1
DIN EN ISO 527-1 for stress–strain measurements
DIN 53504 for stress–strain measurements
DIN EN ISO 1133-1 for measuring the melt flow index
ISO 11443 for viscosity curves by capillary rheometer
EN ISO 6721 for dynamic mechanical analysis
DIN 78004 for TPS (new)
DIN 78005 for TPV (new)

Specifics

Physical basics on polymers

U. Eisele, Introduction to Polymer Physics, Springer Verlag 1990

Rheologie

W-M. Kulicke: Fließverhalten von Stoffen und Stoffgemischen, Hüthig & Wepf Verlag 1986
L.E. Nielsen: Polymer Rheology, M. Decker, New York 1977

Comparison of rubber with other polymers

U. Eisele: KGK 40, No. 6, 1987

Intermitted stress-strain

N. Vennemann, J. Hühndorf, C. Kummerlöwe, P. Schulz: KGK 54, No. 7-8, 2001
Ch.G. Reid, G.K. Cai, H. Tran, N. Vennemann: KGK 57, No. 5, 2004

https://doi.org/10.1515/9783110739848-013

The books of Holden[1] and Drobny[2] present an in-depth view into the structures and manufacturing of TPE: Holden for more structure and properties, Drobny for more structure and processing. The TPE Fibel is a brief introduction of material properties and is the part of injection molding conditions and behavior. The *TPE Magazine*[4] is published every quarter about current and scientific issues in the world of thermoplastic elastomers.

Index

https://doi.org/10.1515/9783110739848-014